Marine Gels

Marine Gels

Editor

Pedro Verdugo

MDPI • Basel • Beijing • Wuhan • Barcelona • Belgrade • Manchester • Tokyo • Cluj • Tianjin

Editor
Pedro Verdugo
University of Washington Friday
Harbor Laboratories
USA

Editorial Office
MDPI
St. Alban-Anlage 66
4052 Basel, Switzerland

This is a reprint of articles from the Special Issue published online in the open access journal *Gels* (ISSN 2310-2861) (available at: https://www.mdpi.com/journal/gels/special_issues/Marine_Gels).

For citation purposes, cite each article independently as indicated on the article page online and as indicated below:

LastName, A.A.; LastName, B.B.; LastName, C.C. Article Title. *Journal Name* **Year**, *Volume Number*, Page Range.

ISBN 978-3-0365-3451-0 (Hbk)
ISBN 978-3-0365-3452-7 (PDF)

Contents

About the Editor vii

Preface to "Marine Gels" ix

Pedro Verdugo
Editorial on Special Issue "Marine Gels"
Reprinted from: *Gels* **2022**, *8*, 150, doi:10.3390/gels8030150 1

Ferenc Horkay
Polyelectrolyte Gels: A Unique Class of Soft Materials
Reprinted from: *Gels* **2021**, *7*, 102, doi:10.3390/gels7030102 3

Dennis A. Hansell and Mónica V. Orellana
Dissolved Organic Matter in the Global Ocean: A Primer
Reprinted from: *Gels* **2021**, *7*, 128, doi:10.3390/gels7030128 21

Pedro Verdugo
Marine Biopolymer Dynamics, Gel Formation, and Carbon Cycling in the Ocean
Reprinted from: *Gels* **2021**, *7*, 136, doi:10.3390/gels7030136 29

Antonietta Quigg, Peter H. Santschi, Adrian Burd, Wei-Chun Chin, Manoj Kamalanathan, Chen Xu and Kai Ziervogel
From Nano-Gels to Marine Snow: A Synthesis of Gel Formation Processes and Modeling Efforts Involved with Particle Flux in the Ocean
Reprinted from: *Gels* **2021**, *7*, 114, doi:10.3390/gels7030114 65

Peter H. Santschi, Wei-Chun Chin, Antonietta Quigg, Chen Xu, Manoj Kamalanathan, Peng Lin and Ruei-Feng Shiu
Marine Gel Interactions with Hydrophilic and Hydrophobic Pollutants
Reprinted from: *Gels* **2021**, *7*, 83, doi:10.3390/gels7030083 79

Toshi Nagata, Yosuke Yamada and Hideki Fukuda
Transparent Exopolymer Particles in Deep Oceans: Synthesis and Future Challenges
Reprinted from: *Gels* **2021**, *7*, 75, doi:10.3390/gels7030075 93

Mónica V. Orellana, Dennis A. Hansell, Patricia A. Matrai and Caroline Leck
Marine Polymer-Gels' Relevance in the Atmosphere as Aerosols and CCN
Reprinted from: *Gels* **2021**, *7*, 185, doi:10.3390/gels7040185 105

About the Editor

Pedro Verdugo Invited Editor of *Marine Gels,* has a background in Chemistry, Medicine, and postdoctoral training in the Thermodynamics of Macromolecules and Bioengineering. With over 92 publications and the editorship of several books, he has had an unorthodox career that includes the discovery that the stochastic transport of eggs in the mammalian oviduct follows Langevinian Gaussian random walk dynamics—in the first section of the tube—and Markovian, Poisson random walk similar to "drunken sailor" formulation, with reflecting and absorbing barrier boundary conditions—in the second section of the oviduct (Biophys. J. 1980). His work on mucus gels brought to light the finding that mucus hydration—a critical feature in mucus function—can be formulated as a characteristic Donnan equilibrium process (Nature 1981, J. Exp. Biol. 1989). His later attention to cellular biophysics focused on intracellular signal transduction in ciliated and secretory cells, including phytoplankton (J. App. Physiol 1980, Biophys. J. 1989, Nature 1998, Biophys. J. 2001), and the discovery that the storage and release of secretory products in exocytosis follows typical Dušek–Tanaka's volume transitions dynamics (Biophys. J 1991). His later work on marine biopolymers brought to light the finding that polymers found in the DOM stock can reversibly associate forming microscopic gels. Their self-assembly follows characteristic second-order kinetics with a thermodynamic yield of about 10% (Nature 1998, Faraday Disc. RSC 2008, Ann. Rev. Mar. Sci 2012, Gels 2021). This discovery has broad significance for the understanding of carbon flux and cycling in the ocean (Wells M.L. Nature 1998) and it was his motivation to accept the appointment as Invited Editor for this Special Issue on *Marine Gels*.

Preface to "Marine Gels"

Despite their utmost critical significance, marine gels are have remained, for a long time, understudied in oceanography research. Compared to Dissolved Organic Matter (DOM), Particulate Organic Matter (POM), including marine gels, compose just a small fraction of the reduced carbon stock present in the ocean. Accordingly, compared to the breadth and strength of publications of the geochemistry of DOM, marine gels have received limited attention. However, we now know that these two pipes of marine carbon flux are not independent, that they are, indeed, interconnected. Because of their polyelectrolyte nature, marine biopolymers—which make the bulk of DOM—can readily self-assemble, forming gels (Nature, 1998). To date, the body of information we possess is still small. Nonetheless, gels might be the crossing bridge that connects the marine living world to the supramolecular world present in the ocean. This assumption was first considered in the pioneering work of Farooq Azam (Science, 1998). Yet, the focus has remained on quantitation, fluctuations, distribution, and the particular chemical structural changes that might make DOM species recalcitrant, to the detriment of studies anon their quaternary structure, shape, size, Z potential, and hydrophobicity, for instance, features required to understand their interactions to form the supramolecular arrays that constitute the matrix of marine gels.

Eighteen years have passed by since the last report on oceanic gel phase at the University of Washington Friday Harbor Labs, which was published in *Marine Chemistry* in 2004. Since then, the understanding of where and how marine gels are formed, their interaction with bacteria and pollutants, and the fascinating discovery that they can function as a nucleating source for cloud formation is attracting the interest of a new cohort of specialists, including colleagues in atmospheric sciences. Yet, this is the beginning of a set of fundamental questions that are likely to have profound significance in marine sciences. This *Marine Gels* Special Issue aims to broaden the community of talent, theory, and technology required to advance our understanding of supramolecular dynamics in the ocean. This issue intends to thoroughly interrogate the complex dynamics of biopolymer interactions in seawater. How susceptible are marine biopolymer dynamics to the consequence of an atmospheric CO2 overload that will certainly not cease to increase in the coming years? This is indeed a fundamental question. Decisions regarding the consequences of catastrophic economic policies are beyond our remit; however, it is in our territory to understand its effects, addressing not only symptomatic particular issues, such as species, regions, and depth, but also the deep macromolecular level where changes can affect every aspect of this complex system. It is important that we understand the fundamental laws that govern the behavior of macro-polyelectrolytes, to investigate the physical rules that underlie marine gel dynamics.

In short, this issue is an effort to alert oceanographers to the remarkable predicting power of polymer physic laws. We also aim to persuade polymer physicists, who constitute the main readership of *Gels*, that the ocean represents an urgent challenge for the survival of our planet, to respond not to the siren appeal from Mars, Jupiter, or distant galaxies, but the urgent call of our own planet. To his end, we invited one of the best young polymer physicists to write a tutorial on polyelectrolyte dynamics, as well as a small but outstanding group of colleagues to outline the many complex challenges that remain to be addressed.

Finally, I want to thank my colleagues, who placed their trust in our editorial effort to bring their outstanding contributions to the press.

Pedro Verdugo
Editor

gels

MDPI

Editorial

Editorial on Special Issue "Marine Gels"

Pedro Verdugo

Friday Harbor Laboratories, Department of Bioengineering, University of Washington, Friday Harbor, WA 98250, USA; verdugo@uw.edu

The ocean is a complex polymer solution. While Marine Sciences comprise a broad set of disciplines, polymer physics has remained largely absent in this front of inquiry. This Special Issue on marine gels is an attempt to alert marine scientists to the powerful predictive tools that physics offers to advance our understanding of the complex polymer dynamics taking place in the ocean. It is also an invitation to polymer physicists to participate in the urgent challenge of exploring the dynamics of marine biopolymers in the world's oceans.

The cosmos, our last frontier, has received much attention, abundant funding, and important progress; however, the ocean, our first frontier—where life was borne and without which life would not exist on earth—remains a challenging field of underfunded riddles. Although explored for centuries, the utmost complexity of the ocean means that it remains one of the least understood subjects in the natural sciences. There is vast and excellent phenomenology at hand including detailed descriptions and correlations among an infinite number of variables; numerous mathematical models; taxonomy of marine living species from bacteria to whales; taxonomy of chemicals and more recent taxonomy of genes, present in seawater. However, the inner works of this gigantic reactor that keeps us alive remain largely a mystery.

The cycling of carbon is the most critical thermodynamic process on our planet, and about half of it takes place in seawater. Understanding how CO_2 is cycled in the ocean is a central issue for the survival of life on earth. While the overall map of carbon cycling is clear, the fundamentals of this process remain mostly unexplored. To remain alive, most living forms—humans in particular—combust organic matter, consume oxygen, and produce CO_2. The reverse crucial cycling of this process results from photosynthesis, half of the global primary production of which is carried out by marine phytoplankton; carbon is fixed by photosynthesis that feeds higher trophic levels. The output of this gigantic photosynthetic reactor yields an annual mean value of ~50 Gt of carbon in the form of biopolymers. While the cell biology of phytoplankton remains obscure, we now know that these unicellulars function as secretory cells, storing biopolymers as condensed-phase polymer networks, and releasing them to seawater via the standard phase transition in exocytosis. The detailed mechanisms of what happens with the near 700 Pg of mostly polymeric-reduced organic material accumulated in seawater, remains hypothetical. What fraction reenters the cycle, consumed by marine biota, and why and what fraction join the recalcitrant organic stock—the most important disposal burial of organic carbon on the planet where discarded molecules remain buried for thousands of years—is still uncertain. Multifactorial modeling can be tweaked to account for multiple outcomes of marine mass transfer but, if untestable, their predictions often turn into formalized speculation.

Despite polymers making up the bulk of organic stock present in the ocean, little of the powerful body of polymer physics, from Florey to Edwards, de Jeans, Dušek, and Tanaka's, has been used to advance our understanding of marine polymer dynamics. Consequently, this Special Issue opens with a short tutorial on polymer networks theory. What follows is the assessment of the global distribution of marine biopolymers, and their chemical, polyelectrolyte, and hydrophobic features. Brief reviews of the application of polymer

Citation: Verdugo, P. Editorial on Special Issue "Marine Gels". *Gels* **2022**, *8*, 150. https://doi.org/10.3390/gels8030150

Received: 22 February 2022
Accepted: 25 February 2022
Published: 28 February 2022

physics theory to explain the reversible association of biopolymers to form marine gels and the critical role of hydrophobic interactions in gel formation ensue. They provide interesting physical–chemical indications of why marine biopolymers either remain in the cycle or enter the largely irreversible recalcitrant burial. How bacteria gain access to metabolizable marine biopolymers is an equally important question, certainly not answered, but addressed here. The role of marine gels in bacterial nutrition, the ion-exchange of heavy metals, and the binding of pollutants is also addressed in this Special Issue.

The deep, dark ocean stores most of the organic stock present in seawater; it presents some of the most intriguing questions in marine science, and has a special place in this set of reports. Finally, closing this set of news from the frontier is the recent and significant discovery that marine gels can be exported to the atmosphere, likely playing a significant role in cloud formation and climate change [1–7].

We, unfortunately, could to retrieve contributions from an important number of invited colleagues, particularly in marine microbiology and marine macrogel dynamics, whose collaboration would have expanded the scope of this issue. However, their work is thoroughly referenced to guide readers to their source.

Funding: This research received no external funding.

Acknowledgments: It is a unique opportunity to have this Gels Special Issue published in print. In my role as invited editor, I want to thank the Multidisciplinary Digital Publishing Institute for broadening the scope of Gels to include this brief outline of some of the critical riddles on the role of gels in the ocean. It is a first step in promoting the collaboration of polymers physicists, microbiologists, geochemists, ecologists, atmospheric scientists, et cetera, in the challenging task of understanding the complex role of marine gels in our planet. In particular, I want to express my deep appreciation to our colleagues Adrian Burd, Wei-Chun Chin, Hideki Fukuda, Dennis Hansell, Ferenc Horkay, Manoj Kamalanathan, Caroline Leck, Peng Lin, Patricia Matrai, Toshi Nagat, Mónica Orellana, Antonietta Quig, Peter Santschi, Ruei-Feng Shiu, Chen Xu, Yosuke Yamada, and Kai Ziervogel, who devoted precious time and entrusted their excellent work to this new interdisciplinary effort.

Conflicts of Interest: The authors declare no conflict of interest.

References

1. Horkay, F. Polyelectrolyte Gels: A Unique Class of Soft Materials. *Gels* **2021**, *7*, 102. [CrossRef] [PubMed]
2. Hansell, D.A.; Orellana, M.V. Dissolved Organic Matter in the Global Ocean: A Primer. *Gels* **2021**, *7*, 128. [CrossRef] [PubMed]
3. Verdugo, P. Marine Biopolymer Dynamics, Gel Formation, and Carbon Cycling in the Ocean. *Gels* **2021**, *7*, 136. [CrossRef] [PubMed]
4. Quigg, A.; Santschi, P.H.; Burd, A.; Chin, W.-C.; Kamalanathan, M.; Xu, C.; Ziervogel, K. From Nano-Gels to Marine Snow: A Synthesis of Gel Formation Processes and Modeling Efforts Involved with Particle Flux in the Ocean. *Gels* **2021**, *7*, 114. [CrossRef]
5. Santschi, P.H.; Chin, W.-C.; Quigg, A.; Xu, C.; Kamalanathan, M.; Lin, P.; Shiu, R.-F. Marine Gel Interactions with Hydrophilic and Hydrophobic Pollutants. *Gels* **2021**, *7*, 83. [CrossRef] [PubMed]
6. Nagata, T.; Yamada, Y.; Fukuda, H. Transparent Exopolymer Particles in Deep Oceans: Synthesis and Future Challenges. *Gels* **2021**, *7*, 75. [CrossRef]
7. Orellana, M.V.; Hansell, D.A.; Matrai, P.A.; Leck, C. Marine Polymer-Gels' Relevance in the Atmosphere as Aerosols and CCN. *Gels* **2021**, *7*, 185. [CrossRef]

gels

Tutorial

Polyelectrolyte Gels: A Unique Class of Soft Materials

Ferenc Horkay

Section on Quantitative Imaging and Tissue Sciences, Eunice Kennedy Shriver National Institute of Child Health and Human Development, National Institutes of Health, Bethesda, MD 20892, USA; horkayf@mail.nih.gov

Abstract: The objective of this article is to introduce the readers to the field of polyelectrolyte gels. These materials are common in living systems and have great importance in many biomedical and industrial applications. In the first part of this paper, we briefly review some characteristic properties of polymer gels with an emphasis on the unique features of this type of soft material. Unsolved problems and possible future research directions are highlighted. In the second part, we focus on the typical behavior of polyelectrolyte gels. Many biological materials (e.g., tissues) are charged (mainly anionic) polyelectrolyte gels. Examples are shown to illustrate the effect of counter-ions on the osmotic swelling behavior and the kinetics of the swelling of model polyelectrolyte gels. These systems exhibit a volume transition as the concentration of higher valence counter-ions is gradually increased in the equilibrium bath. A hierarchy is established in the interaction strength between the cations and charged polymer molecules according to the chemical group to which the ions belong. The swelling kinetics of sodium polyacrylate hydrogels is investigated in NaCl solutions and in solutions containing both NaCl and $CaCl_2$. In the presence of higher valence counter-ions, the swelling/shrinking behavior of these gels is governed by the diffusion of free ions in the swollen network, the ion exchange process and the coexistence of swollen and collapsed states.

Keywords: polyelectrolyte; gel; swelling; ions; volume phase transition; osmotic swelling pressure; elastic modulus; swelling kinetics

Citation: Horkay, F. Polyelectrolyte Gels: A Unique Class of Soft Materials. *Gels* **2021**, *7*, 102. https://doi.org/10.3390/gels7030102

Academic Editor: Pedro Verdugo

Received: 1 June 2021
Accepted: 21 July 2021
Published: 24 July 2021

1. Introduction

It is often claimed that the gel state of a material is easier to recognize than to define. A widely accepted phenomenological definition is that a gel is a soft, solid or solid-like material of two or more components, one of which is a liquid, present in a substantial quantity [1,2]. From a rheological point of view, gels are characterized by the storage modulus, G′, which exhibits a plateau extending to times of the order of seconds, and by a loss modulus, G″, which is much smaller than the storage modulus in the plateau region. This definition is consistent with that of Ferry [3]. The term gel has also been used for systems that do not contain a liquid (e.g., vulcanized rubber or dried silica gel).

Based on the mechanism of the cross-linking process, gels can be classified as physical or chemical networks [4,5]. Physical cross-links may arise from hydrogen bonding, hydrophobic interactions, inter-chain entanglements and crystallite formation. Physical cross-linking produces reversible gels. Although physical cross-links are not permanent, they are sufficiently strong to tie the polymer chains together.

Chemically cross-linked gels can be formed by various chemical processes such as free radical polymerization, electromagnetic radiation (light, gamma, X-ray or electron beam) and chain or step-growth polymerization. In all cases, covalent cross-links alter the chemical structure of the polymer and have significant consequences on the physical properties of the system at both molecular and supramolecular levels. Cross-linking renders the polymer insoluble, and the type and extent of cross-linking influence important network properties, such as swelling, elastic and transport properties. In a given solvent, the degree of swelling depends on the cross-link density of the network and the interaction between the polymer and solvent. Uncross-linked polymers can be diluted infinitely. Cross-links prevent infinite swelling because the osmotic mixing pressure Π_{mix}, which is the driving force of swelling, is counter-balanced by the elastic pressure Π_{el} generated by the cross-links. At equilibrium, the swollen network coexists with the pure solvent.

Typical examples of physical gels are those formed by the cooling of solutions of biological or synthetic polymers (e.g., gelatin, agarose, polyvinyl alcohol). Typical chemical gels are dextrane gels (e.g., Sephadex), polyvinylalcohol gels, polystryrene gels, etc. Examples of biological gels include cartilage or fibrin clots formed by polymerization of fibrinogen monomers through enzymatic reactions. Silica gel is a typical example of an inorganic gel. The common feature of these materials is that they are soft, solid or solid-like and contain a liquid. Hydrogels are networks of hydrophilic polymers swollen in water. In organogels, the polymer network is hydrophobic, and the liquid is an organic solvent (e.g., toluene).

2. Effect of the Environment on the Swelling of Gels

The properties (e.g., swelling degree, elastic modulus) of many hydrogels are sensitive to changes in the environmental conditions (e.g., pH, ionic strength, solvent composition, temperature). For example, in pH- and ion-responsive gels, a volume transition can be induced by changing the ionization of the polyelectrolyte chains. By decreasing the degree of ionization, the electrostatic repulsion between the charged groups on the polymer chains is reduced, which ultimately leads to the collapse of the swollen network. In temperature-sensitive systems, the strength of polymer–solvent contacts varies relative to the polymer–polymer contacts, and the gel undergoes a volume transition. Light and the electric field can also induce changes in the swelling degree of gels. In the former, the volume change is due to the temperature increase caused by photosensitive groups (chromophores) which absorb light and dissipate heat. In electrosensitive gels, the applied electric field attracts mobile ions to the electrodes, and the hydrogel swells (or shrinks) at the cathode and the anode.

Temperature-sensitive hydrogels contain both hydrophilic and hydrophobic monomers. At low temperature, hydrogen bonding between the polymer and water molecules leads to dissolution in water. However, when the temperature exceeds a critical temperature (lower critical temperature, LCST), the hydrogen bonds break down, and phase separation occurs. Varying the relative amounts of hydrophilic and hydrophobic monomers can alter the LCST of hydrogels.

Stimuli-responsive hydrogels are widely used in drug delivery devices to deliver drugs to a specific site in the body. These systems are frequently called smart or intelligent gels because the fast response to external stimuli is a typical feature of living systems [6–9]. Poly(N-isopropylacrylamide) (PNIPAM) is the most studied thermosensitive hydrogel in drug delivery applications. This is due to the ability of PNIPAM to squeeze out the absorbed drug when the temperature is near that of the human body. Glucose sensors are used in insulin delivery systems [10]. Another important application of smart gels is scaffolds for tissue engineering because they are capable of releasing cells in response to a stimulus [11,12].

3. Modeling of Polymer Gels

Modeling the response of gels to changes in the environmental conditions is a complex task that can be attempted at different length and time scales. The classical theory of Flory and Huggins describes gels in terms of osmotic and elastic forces, which define the degree of swelling [4]. High polymer–solvent affinity, i.e., a strong interaction among the polymer and solvent molecules, leads to a large osmotic pressure. The elastic contribution can be estimated from the theory of rubber elasticity [4,13]. In the case of charged polymer networks, electrostatic interactions and counter-ion osmotic pressure may also play a significant role [4].

Computer simulations provide valuable information on the behavior of polymer gels. There are various simulation methods to study polymer gels (e.g., Monte Carlo simulation, molecular dynamics, Brownian dynamics) [14–19]. The choice of the method depends on the focus of the study. The first step before conducting the simulation is to define the level of complexity required to capture the features of the system. It is common to use coarse graining to model polymer chains. The next step is to introduce intermolecular and intramolecular interactions and generate the initial configuration. Then, this configuration is relaxed, and the properties of the system are monitored until an equilibrium is attained. A typical approach is to study the dynamics of only a limited number of particles, while other particles (e.g., solvent molecules) are viewed as a continuous phase. This implicit solvent approach has the advantage of being computationally less expensive, thus making it possible to simulate larger systems over longer time scales.

Over the past decade, computer simulation has proved to be a uniquely powerful tool in investigating polymer gels. It allows constructing near-perfect model networks with a well-defined topology. The network topology can be systematically varied from ideal to more realistic systems by introducing a controlled quantity of structural defects (entanglements, loops, dangling ends, etc.). In real gels, imperfections formed during the cross-linking process are always present. Computer simulation makes it possible to determine the effect of inhomogeneities on the material properties (e.g., mechanical strength). Earlier simulation studies largely focused on the dynamic properties of these systems, e.g., kinetics of cross-linking. Much less attempt has been made to simulate the swelling of gels that may require a very long time to reach an equilibrium. Future studies should focus, among others, on the effect of the size and structure of solvent molecules, the influence of the temperature on the interaction potentials and other molecular details of the network structure, such as network functionality and structural irregularities.

4. Polyelectrolyte Gels

Polyelectrolytes are macromolecules containing ionizable groups, which, in aqueous solutions, dissociate and release counter-ions into the solution. These charged macromolecules play important roles in various processes in living systems such as DNA condensation, nerve excitation and load bearing of cartilage. Typical examples of biological polyelectrolytes are nucleic acids, proteins and proteoglycans.

In general, in polyelectrolyte systems, electrostatic interactions between the polymer molecules and ions lead to a very rich behavior, which differs in many aspects from the behavior of neutral polymers. Polyelectrolyte gels contain charged groups on the cross-linked polymer chains, and ions in the swelling liquid [20–22]. They exhibit unique properties because the effects of ions on polyelectrolyte molecules are not only short range but, due to the connected structure of the polymer chains, also long range. Due to the presence of ionized groups and mobile ions, polyelectrolyte gels are sensitive to external stimuli such as changes in pH, ionic interactions and temperature, and they may exhibit a volume transition in response to these changes [23–27].

Polyelectrolyte gels have a large number of ionizable groups. They respond to the change in the pH in the surrounding liquid by either gaining or losing protons. In basic environments, anionic polyelectrolytes are deprotonated, and the strong electrostatic repulsion among the anionic chains leads to gel swelling. In acidic environments, the anionic polymer is protonated, resulting in a decrease in the charge density, and the gel collapses. Cationic polyelectrolytes exhibit the opposite behavior; they are ionized and swell in acidic environments and collapse in basic environments. There are polymers which contain both anionic and cationic groups. Such amphiphilic hydrogels swell in both acidic and basic environments. The pH dependence of the behavior of different charged hydrogels is illustrated in Scheme 1.

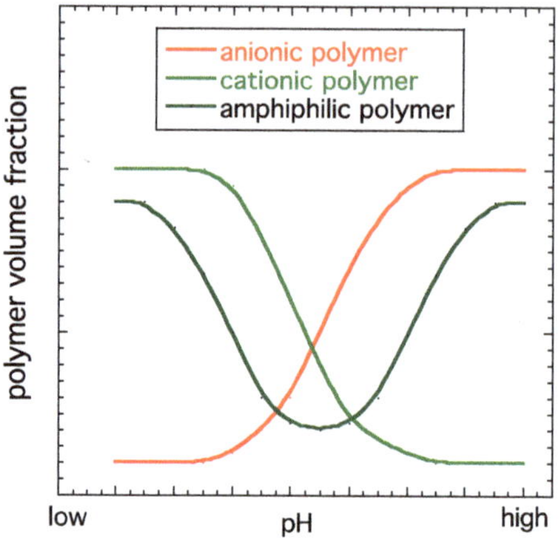

Scheme 1. Schematic representation of the variation in the polymer volume fraction as a function of the pH in different types of polyelectrolyte gels. Anionic gels swell in basic solutions (high pH), and cationic gels swell in acidic solutions (low pH). Amphiphilic gels swell in both low- and high-pH solutions.

Several theoretical models have been proposed to describe volume transitions in polyelectrolyte gels [28]. Katchalsky [29] extended the Flory–Rehner model developed for neutral polymer gels to describe the swelling of polyelectrolyte gels. It was shown that (i) the swelling of polyelectrolyte gels was governed by the balance between the elastic free energy of the cross-linked polymer chains and the osmotic pressure of the charged polymer and counter-ions in the gel, and (ii) a small variation in the salt concentration in the equilibrium solution could induce a volume transition in the polyelectrolyte gel. The condition of electroneutrality insured that the ion concentrations in the gel satisfied the Donnan equilibrium condition.

The discontinuous volume transition in gels was first described by Dusek and Patterson [30] based on the analogy of the coil–globule transition of polymer chains in solution. The high swelling ability of polyelectrolyte gels reflects the electrostatic repulsion between the charged groups on the polymer chains. The addition of high-valence counter-ions screens the electrostatic interaction and leads to phase separation. Monovalent counter-ions are gradually replaced by high-valence counter-ions, and the osmotic pressure is reduced. At a critical threshold concentration of the multivalent ions, the osmotic pressure vanishes, and phase separation takes place.

Less attention has been paid to the kinetics of polyelectrolyte gel swelling. Gel swelling/shrinking involves the motion of both polymer and solvent molecules. The response time to changes in the environment (ion concentration and composition, temperature, pH, etc.) strongly depends on the actual geometry of the gel. Swelling kinetics measurements provide quantitative information on the collective diffusion coefficient.

5. Gels in Living Systems

Biopolymers are naturally occurring macromolecules which are essential components of all living systems [31,32]. Many biopolymers are polyelectrolytes, i.e., charged macromolecules. Understanding the behavior of polyelectrolytes is one of the most challenging problems in polymer science [33–35]. The properties of these molecules reflect chain connectivity, electrostatic effects and various molecular interactions over multiple length scales. Owing to the complexity of these interactions, the progress in this field has been slow despite the importance of polyelectrolytes both in biology and materials science.

In general, biological tissues are highly swollen polyelectrolyte gels. Many important tissue properties originate from the polyelectrolyte nature of their constituents. For example, articular cartilage consists of an extracellular matrix (ECM) containing negatively charged aggrecan/hyaluronic acid complexes embedded in a collagen matrix. In other biological systems, e.g., in the nervous system, Na^+, K^+ and Ca^{2+} ions regulate the excitability of neurons. Intracellular Ca^{2+} ions play an important role in a variety of physiological processes such as muscle contraction, hormone secretion, synaptic transmission and gene expression.

Despite the many recent efforts in the field of polyelectrolytes, no satisfactory theoretical model exists that describes the behavior of these systems and provides quantitative agreement with the experimental findings. Problems such as counter-ion condensation, coupling between small ions and macro-ions and the effect of counter-ions on chain stiffness, which are necessary to understand the behavior of polyelectrolyte gels, have not yet been fully resolved. Polyelectrolyte gels have a great potential not only for designing new functional biomaterials (e.g., artificial muscle, biosensors) but also for understanding the principles of complex biological systems such as cartilage. Progress in the field requires an interdisciplinary effort to accomplish a better understanding of the structure and interactions of polyelectrolyte systems over multiple length and time scales.

6. Marine Microgels

Marine microgels are polymer networks formed by spontaneous assembly of biopolymers in the ocean with seawater entrapped in the swollen network [36]. Better understanding the behavior of these gels is particularly important because the ocean plays a critical role in global carbon cycling: it handles half of the global primary production of reduced organic carbon. As dissolved organic carbon is available for marine microorganisms (e.g., bacteria), the ocean is a huge repository of carbon [37–39]. Marine microgels exist in the ocean, i.e., an environment in which both mono- and multivalent counter-ions are present. The important role of counter-ions in the swelling of various polyelectrolyte gels was recognized a long time ago. Polymer physics provides valuable insights into the behavior of polyelectrolyte gels, whose knowledge is essential to understanding the role of marine biopolymers in carbon cycling.

In the ocean, carboxylic acids are the most common negatively charged residues present in dissolved organic matter. Carboxylic acids contain a hydrogen ion that other ions (e.g., Na^+, Ca^{2+}, Mg^{2+}) can replace. Counter-ions interact with oppositely charged sites. Polyanions with few available polyanionic sites are likely to form relatively unstable assemblies. An important feature of electrostatic interactions in polyelectrolytes is that the probability of forming links is proportional to the second power of the valence of the counter-ion [40]. For example, Fe^{3+} can induce self-assembly of the organic components in seawater at much lower concentrations and in shorter times than divalent counter-ions. The characteristic coagulating effect of low concentrations of Al^{3+} salts in seawater is another outcome that can be explained by the strong effect of high-valence polyions on the assemblies of dissolved carbon nutrients [41].

In this article, we use a didactic approach. First, we describe the fundamental properties of negatively charged polyelectrolyte gels. Then, experimental results are shown both for a synthetic (sodium polyacrylate, PA) and a biopolymer gel (DNA) swollen in mono- and multivalent salt solutions. The advantage of performing measurements on model polyelectrolyte hydrogels is that their structure is well defined (unlike the structure of most natural gels), which is essential for conducting a systematic and quantitative study.

The following sections of this paper are organized as follows: After presenting the brief theoretical framework, describing the equilibrium swelling behavior of polyelectrolyte gels and the kinetics of gel swelling, the thermodynamic analysis of swelling equilibrium measurements conducted on gels swollen in solutions containing monovalent, divalent and trivalent salts is discussed. This is followed by the analysis of the swelling kinetics measurements. We compare the swelling kinetics of PA gels in salt-free water and in solutions containing both mono- and divalent cations. The results clearly indicate the important role of both ion exchange and ion diffusion in the development of complex structures consisting of coexisting shrunken and swollen regions. Finally, the main results are summarized in the conclusions.

We believe that understanding the behavior of solutions of charged biopolymers in near-physiological salt solutions will shed light on the mechanism of structure formation in various biological systems (e.g., organization of structural components of the cell and extracellular matrix). Although the experimental work presented here was conducted on model hydrogels, it is reasonable to assume that the response of these gels to changes in the ionic environment is similar to other polyelectrolyte gels, including marine microgels.

7. Theory

7.1. Swelling of Polyelectrolyte Gels

Polyelectrolyte gels are polymer networks in which the charged sites are fixed on the polymer chains, and the counter-ions in the surrounding liquid ensure electroneutrality. Due to the repulsive interaction between identically charged groups on the macromolecules, the dry network swells by absorbing solvent (e.g., water) molecules. The amount of water absorbed by a typical polyelectrolyte network can exceed 1000 times the weight of the dry polymer. However, the polymer network cannot be dissolved because of the presence of permanent cross-links. Changing the environment of the gel (salt concentration, pH, etc.) affects its swelling degree. As discussed earlier, with an increasing salt concentration, the counter-ions gradually screen the electrostatic repulsion between the charged groups on the polymer chains, and above a critical threshold, the salt concentration leads to the collapse of the gel. This effect can be reversed by removing the salt from the collapsed gel. The swelling process is illustrated in Scheme 2.

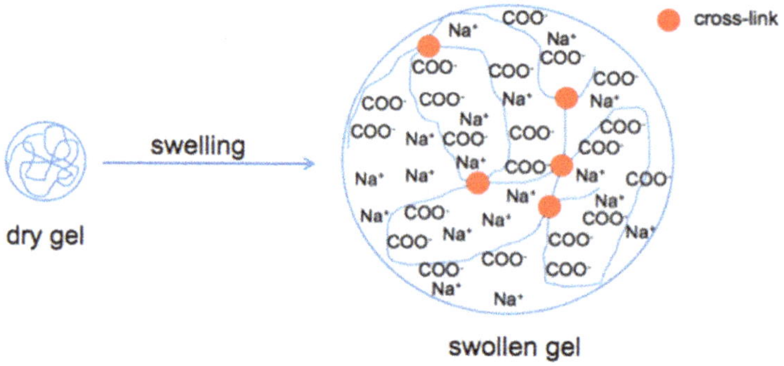

Scheme 2. Swelling of a polyelectrolyte gel. Water uptake is driven by the osmotic pressure difference between the polymer network and the surrounding solution.

The stability of a polyelectrolyte gel is the result of a delicate balance between several competing thermodynamic forces. At equilibrium, the free energy, ΔF_{tot}, of the swollen network reaches a minimum. In the case of neutral polymer gels, ΔF_{tot} is the sum of the free energy of elastic deformation of the network chains, ΔFel, and the free energy of the mixing of polymer and solvent molecules, ΔFmix [4,13,42]. In the case of polyelectrolytes, however, there is an additional term, ΔFion, due to the presence of the counter-ions (Donnan contribution). Assuming that these terms are independent, we can write [4]

$$\Delta F_{tot} = \Delta F_{el} + \Delta F_{mix} + \Delta F_{ion} \quad (1)$$

In an osmotic swelling experiment, the derivatives of the free energy components are measured, i.e.,

$$\Pi_{tot} = -\partial(\Delta F_{tot}/V_1)/\partial n_1 = \Pi_{el} + \Pi_{mix} + \Pi_{ion} \quad (2)$$

where Π_{tot} is the swelling pressure of the gel, Π_{el}, Π_{mix} and Π_{ion} are the elastic, mixing and ionic contributions of Π_{tot}, V_1 is the molar volume of the solvent, and n_1 is the number of moles of the solvent.

For networks made of flexible polymer chains, the elastic contribution can be estimated from the theory of rubber elasticity [13,42]:

$$\Pi_{el} = -ART\nu\ \varphi^{1/3} = -G \quad (3)$$

where ν is the concentration of the elastic chains, φ is the volume fraction of the polymer, R is the gas constant, and T is the absolute temperature. The constant, A, depends on the functionality of the junctions and the topology of the network. Π_{el} can be expressed by the shear modulus, G, of the gel.

The osmotic mixing pressure, Π_{mix}, can be given by the Flory–Huggins expression:

$$\Pi_{mix} = -RT/V_1\ [\ln(1-\varphi) + \varphi + \chi\ \varphi^2 + \chi_1\varphi^3] \quad (4)$$

where χ_0 and χ_1 are constants (interaction parameters).

The ionic contribution is due to the difference in mobile ion concentrations inside and outside the polyelectrolyte gel [4], which gives rise to an osmotic pressure difference, Π_{ion}, between the gel and the equilibrium solution. According to the theory of Donnan,

$$\Pi_{ion} = RT \sum_{j=1}^{N} (cj^{gel} - cj^{sol}) \quad (5)$$

where cj_{gel} and cj^{sol} represent the concentrations of the ions in the gel and in the equilibrium solution, and N is the number of mobile ions in the system.

Recent experimental studies [43–47], as well as molecular dynamics simulations [48], indicated that in the presence of added salt, the Donnan contribution is very small, i.e., the first two terms of Equation (2) provide a satisfactory description of the osmotic properties of polyelectrolyte gels.

7.2. Kinetics of Gel Swelling

The kinetic theory of gel swelling is based on the concept of cooperative diffusion of the network polymer in a continuous medium [49,50]. According to the theory of Tanaka,

$$u(r,\ t)/u(r,\ \infty) = \sum_i B_i\ \exp\left[-t/\tau_i\right] \quad (6)$$

where u(r, t) is the displacement vector, t is the swelling time, and τ_i is the relaxation time. The numerical factor B_i depends on the geometry of the gel and the ratio of the shear modulus G over the longitudinal osmotic modulus M_{os}. This theory views gel swelling as a combination of pure diffusion and pure shear relaxation processes. The important role of the shear modulus is to keep the system in shape. Since during a shear relaxation process, there is no relative motion and, consequently, no friction between the polymer and the solvent, the gel can instantaneously adjust its shape, thereby minimizing nonisotropic deformation.

At a large t, the slowest term of Equation (6) dominates. For spherical gels, the theory predicts

$$d_t = d_\infty + [d_0 - d_\infty]\ B_1\ \exp(-t/\tau_1)\ (t > \tau_1) \quad (7)$$

where d_t is the diameter of the gel at time t, d_0 and d_∞ are the initial and final (equilibrium) diameters, and τ_1 is the relaxation time of the slowest mode in the swelling process.

Since gel swelling is a diffusion-controlled process, the rate of swelling (or shrinking) depends on the size of the sample. The collective diffusion coefficient D_c is given by

$$D_c = M_{os}/f = a^2/(\beta_1{}^2\tau_1) \quad (8)$$

where f is the friction coefficient, a is the radius of the gel, and β_1 is the function of G/M_{os}. This collective mode of diffusion, which is distinct from the translational motion either of the individual solvent molecules or the polymer chains, governs the rate at which polymer and solvent molecules mutually exchange positions upon swelling or deswelling. Specifically, D_c defines the rate at which the solvent enters or leaves each elementary fluctuating volume. The size of this volume depends on the thermodynamic conditions, notably the polymer concentration and the salt content. Thus, the value of D_c reflects changes in the ionic environment. For neutral polymer gels, D_c is typically in the order of 10^{-6}–10^{-7} cm^2/s.

8. Results and Discussion

In this section, we show typical experimental findings for model polyelectrolyte gels. First, we focus on the effect of counter-ion valence on the macroscopic swelling behavior of sodium polyacrylate (PA) and DNA gels. We separate the elastic and mixing contributions of the swelling pressure and investigate the variation in these components with the ion valence and the ion concentration.

8.1. Effect of Salts on the Swelling and Osmotic Behavior of Model Polyelectrolyte Gels

Figure 1 shows the dependence of the polymer volume fraction φ of PA hydrogels as a function of the salt concentration, c_{salt}, in solutions of monovalent (Figure 1a) and multivalent salts (Figure 1b). Gel swelling is the greatest in salt-free solutions. The general trend is that the salt addition induces gel contraction because ions screen the repulsive electrostatic interactions between the charged groups on the polymer chains. However, the valence of counter-ions has a significant effect on the swelling behavior. In monovalent salt solutions (LiCl, NaCl, KCl, RbCl, CsCl), the polymer concentration increases from $\varphi \approx 0.001$ to $\varphi \approx 0.01$. It can also be seen that the effect of different divalent ions is similar, indicating that the ion valence, rather than the chemical nature of the cation, determines the swelling of the present anionic PA gels.

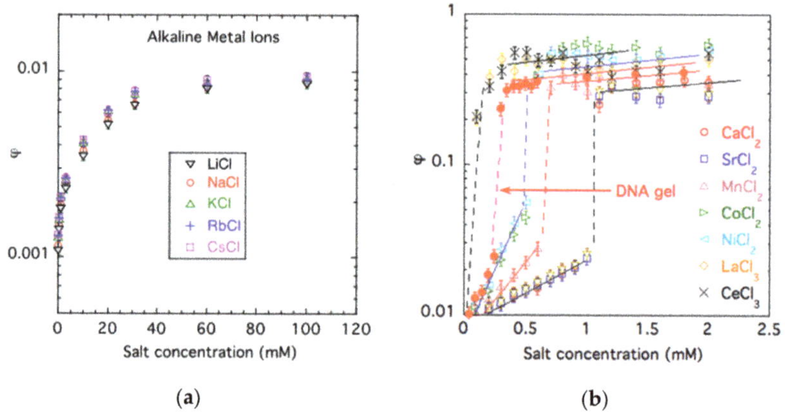

Figure 1. Effect of different monovalent (**a**) and multivalent salts (**b**) on the volume fraction φ of PA gels. Filled symbols show the dependence for a DNA gel. In (**b**), measurements were conducted in solutions containing 40 mM NaCl. Curves are a guide for the eye.

In the presence of divalent counter-ions (Ca^{2+}, Sr^{2+}, Mn^{2+}, Co^{2+}, Ni^{2+}) at a 'critical concentration' (or a critical ratio of divalent to monovalent cations), a reversible volume transition occurs. The sharp volume change with an increasing concentration of divalent counter-ions indicates that the transition is a highly cooperative process. Figure 1b also shows the data for a weakly cross-linked DNA gel swollen in $CaCl_2$ solutions, which exhibits a qualitatively similar behavior to the PA gels, although the numerical values are different.

In the PA gel, the volume transition occurs at approximately 1 mM $CaCl_2$ concentration, while in the DNA gel, it takes place at a lower calcium ion content (≈0.25 mM $CaCl_2$).

Trivalent cations shift the transition concentration toward lower salt concentrations. We note that the present trivalent counter-ions (Ce^{3+} and La^{3+}) practically bind irreversibly to the polyanion.

A volume change reflects the interplay between two main effects: attractive interactions among the polymer segments, which tend to shrink the gel, and repulsion of similarly charged species (either charged units of the polymer network or mobile ions). Consequently, any change in the mixing and elastic free energy contributions affects the polymer concentration.

To reveal the effect of added salts on the equilibrium swelling of gels, the elastic and mixing contributions of the free energy should be separately investigated. First, we quantify the effect of ions on the elastic modulus of the gels. Then, we focus on the osmotic mixing component by conducting swelling pressure measurements on gels containing different salts. We compare the osmotic results for two entirely different gel systems: PA and DNA gels.

It is often assumed that divalent cations form bridges between the charged groups on the polyelectrolyte network. The formation of ion bridges is expected to increase the apparent cross-link density and thus the elastic (shear) modulus of the gel. In Figure 2, the dependence of the shear modulus on the polymer volume fraction of PA gels measured in $CaCl_2$ and $CoCl_2$ solutions is shown. In $CaCl_2$ solutions, all data points fall on a single curve, indicating that G is a function of the polymer concentration only, i.e., the concentration of Ca^{2+} ions does not modify the 'effective' cross-link density of the gel. At low and moderate volume fractions, G varies according to the power law prediction of the theory of rubber elasticity, i.e.,

$$G = G_o \varphi^{1/3} \tag{9}$$

where G_o is a constant (G_o = ARTν). Deviation from the theoretical dependence is observed only in the most swollen gels (without an added salt), where the shear modulus increases with the decreasing volume fraction. In such highly swollen gels, the finite extensibility of the network chains dominates, and the elastic response can no longer be described by the Gaussian elasticity theory. A similar upturn in the elastic modulus at high swelling degrees was observed previously in ionized acrylamide-sodium acrylate copolymer gels [51,52].

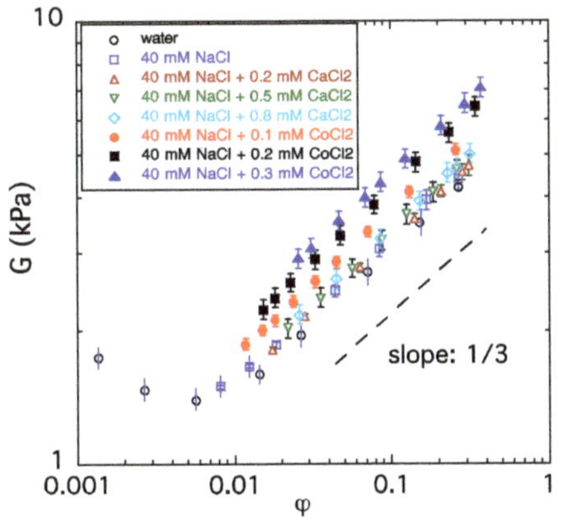

Figure 2. Variation in the shear modulus of PA gels swollen in different salt solutions and water.

In general, alkaline earth metal ions cause gel contraction but do not form additional cross-links. Figure 2 also shows that when Ca^{2+} ions are replaced by Co^{2+} ions, the value of G increases with the increasing $CoCl_2$ concentration (filled symbols). A probable explanation of the increase in the elastic modulus is complex formation between the Co^{2+} ions and the polyacrylate anion.

Figure 3 shows the total swelling pressure of a PA gel, Π_{tot}, and its elastic Π_{el} and mixing Π_{mix} components at a constant ion concentration (c_{salt}: 40 mM NaCl solution). Π_{tot} was determined from osmotic stress measurements, Π_{el} was estimated from the shear modulus ($\Pi_{el} = -G$) and Π_{mix} was calculated using the relationship $\Pi_{mix} = \Pi_{tot} + G$. Both Π_{tot} and Π_{mix} increase while Π_{el} decreases with the increasing polymer volume fraction. The continuous curve is the least squares fit of Equation (4) to the Π_{mix} data, which yields, for the interaction parameters, $\chi_0 = 0.448 \pm 0.001$ and $\chi_1 = 0.21 \pm 0.01$.

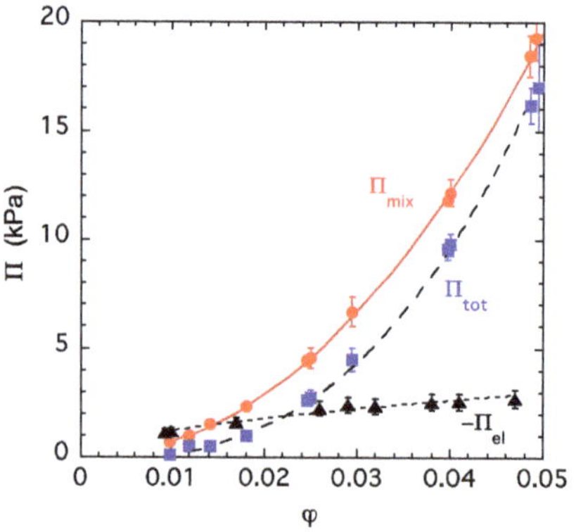

Figure 3. Variation in the swelling pressure Π_{tot} and its osmotic Π_{osm} and elastic Π_{el} components with the volume fraction of the polymer in a PA gel swollen in 40 mM NaCl solution.

It is reasonable to assume that specific interactions between the counter-ions and the carboxylate groups may modify the mixing contribution of the network free energy. The dependence of Π_{mix} as a function of φ is shown in Figure 4a for PA gels swollen in 40 mM NaCl solutions containing different amounts of $CaCl_2$ or $CoCl_2$. It can be seen that both the chemical type and the concentration of the cations affect the mixing pressure. Π_{mix} (i) decreases with the increasing concentration of divalent cations and (ii) depends on the chemical type of the cation; the effect of Co^{2+} ions is qualitatively similar but significantly greater than that of the Ca^{2+} ions. The lines through the experimental data points are least squares fits to Equation (4). As shown in the inset, the addition of divalent counter-ions causes a very weak increase in the value of χ over the entire concentration range explored, while χ_1 strongly increases as the divalent ion concentration increases and thereafter exhibits a slow increase. Figure 4b shows similar data for a DNA gel swollen in 40 mM NaCl solution with different $CaCl_2$ contents. Again, Π_{mix} decreases with increasing Ca^{2+} concentration. The variation in χ and χ_1 with the $CaCl_2$ concentration is similar in the DNA and PA gels, indicating that gel swelling is primarily governed by the electrostatic interactions in the system, and the role of the chemistry of the polymer chains is less important.

In summary, based on the effect of ions on the apparent cross-link density and the osmotic mixing contribution, a hierarchy can be established in the interaction strength according to the chemical group to which the ions belong. Alkali metal ions and alkaline earth metal ions do not considerably affect the elastic properties of PA hydrogels. The effects of alkaline earth metal salts can be attributed to modification of the mixing free energy. Experimental data indicate that the addition of $CaCl_2$ has little effect on χ but significantly increases χ_1. Transition metal ions (Co^{2+}, Ni^{2+}) form complexes with the polyanion, which can be interpreted as additional cross-links, as indicated by the increase in the elastic modulus. In these gels, cations affect both the elastic and mixing free energy terms. Trivalent cations (La^{3+} and Ce^{3+}) practically bind irreversibly to the polyanion.

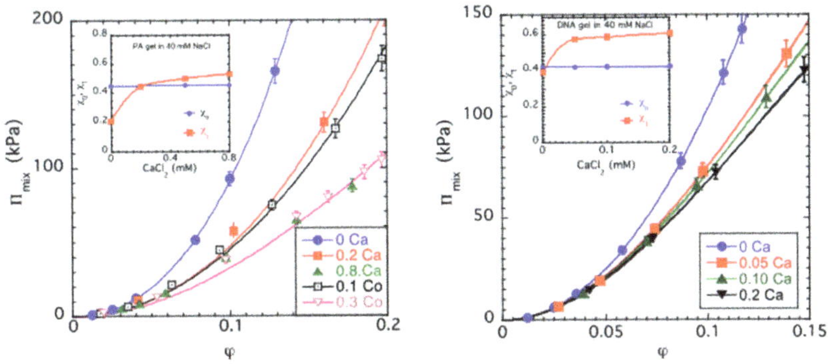

(**a**) (**b**)

Figure 4. Mixing pressure, Π_{mix}, vs. polymer volume fraction, φ plots for PA (**a**) and DNA gels (**b**) in different salt solutions. The continuous lines show the fits of Equation (4) to the experimental data. Insets: variation in χ and χ_1 as a function of the $CaCl_2$ concentration in PA and DNA gels swollen in 40 mM NaCl.

Gel swelling is primarily governed by the effect of ions on the electrostatic interactions among the charged groups of the polymer network. With the increasing concentration of the added salt, the repulsive interaction is gradually screened, and above a critical concentration of the counter-ions, a sudden volume change takes place. The decisive factor in the volume transition is the ion valence, while the chemical type of the ions and the chemical composition of the polymer network are of secondary importance.

8.2. Kinetics of Polyelectrolyte Gel Swelling in Salt Solutions

In Figure 5, the swelling kinetics data measured in 40 mM NaCl solution for spherical PA gels of different sizes are shown [53]. All samples exhibit a qualitatively similar behavior: the gel size monotonically increases as a function of time and asymptotically approaches a plateau. As expected, smaller gels swell faster. The inset in Figure 5 illustrates the variation in the reduced gel diameter d_t/d_o with the reduced time t/τ_1 for the same gels, as in the main figure (upper curves), and for gels swollen in 100 mM NaCl solutions (lower curves). At each salt concentration, the data points fall on a master curve. The height of the plateau region, corresponding to the equilibrium concentration of the fully swollen gels, decreases with the increasing salt concentration.

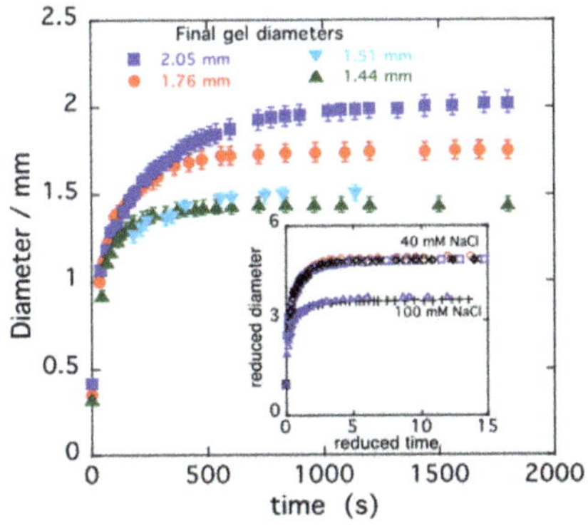

Figure 5. Swelling kinetics data of PA gels in 40 mM NaCl solution. Inset: reduced gel diameter d/d_0 vs. reduced time t/τ_1 for PA gels swollen in 40 mM NaCl and 100 mM NaCl solutions.

The analysis of the swelling kinetics data was performed using the linearized form of Equation (7):

$$Y = \ln\left[(d_\infty - d_t)/(d_\infty - d_0)\right] = \ln B_1 - t/\tau_1 \quad (10)$$

From the intercept of the long-time linear extrapolation of the logarithmic plot, and from the slope of the straight line, B_1 and τ_1, respectively, can be determined. Once B_1 is known, β can be found since both B_1 and β depend on G/M_{os}, and this dependence has been established numerically for spheres in the literature [50].

In Figure 6, the quantity Y is plotted as a function of time for spherical gels with different initial diameters. At higher values of t, all gels exhibit a linear behavior, as predicted by Equation (10), in agreement with the expectation that the relaxation time decreases with decreasing gel size. B_1 is, however, the same for the four gel samples shown in Figure 5, indicating that the ratio of the moduli G/M_{os} is independent of the gel size. The values of τ_1 and B_1 obtained from the fits were used to calculate D_c.

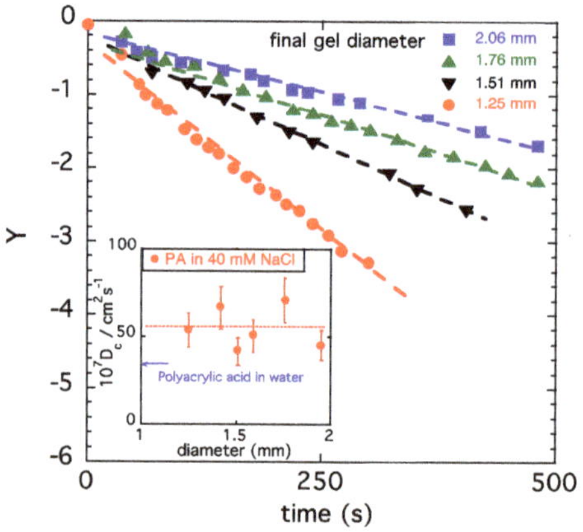

Figure 6. Variation in Y in Equation (10) as a function of the swelling time for PA gels swollen in 40 mM NaCl solution. Dashed curves are least squares fits to Equation (10). Inset: collective diffusion coefficient D_c of different hydrogels.

In the inset, the cooperative diffusion coefficient D_c is presented for PA gels of different sizes. In Figure 6, the value of D_c reported for a polyacrylic acid gel [51] measured in salt-free water by dynamic light scattering is also displayed. The present result shows that D_c in 40 mM NaCl solution exceeds the value measured in pure water by approximately 30%.

The presence of divalent cations in the solution affects the swelling/shrinking kinetics of PA hydrogels. Figure 7 shows the variation in the diameter of a PA gel after the addition of 10 mM $CaCl_2$ to the surrounding 40 mM NaCl solution. In the shrinking process, three different stages can be clearly distinguished. In the first stage, the swelling degree decreases slowly, practically linearly with the time. In the course of this process, the Ca^{2+} concentration progressively increases from the surface of the gel to the center. At a certain Ca^{2+} concentration, a volume transition occurs. First a collapsed layer is formed on the surface, which expands as the Ca^{2+} front moves towards the center. In this stage, the collapsed network coexists with the swollen gel.

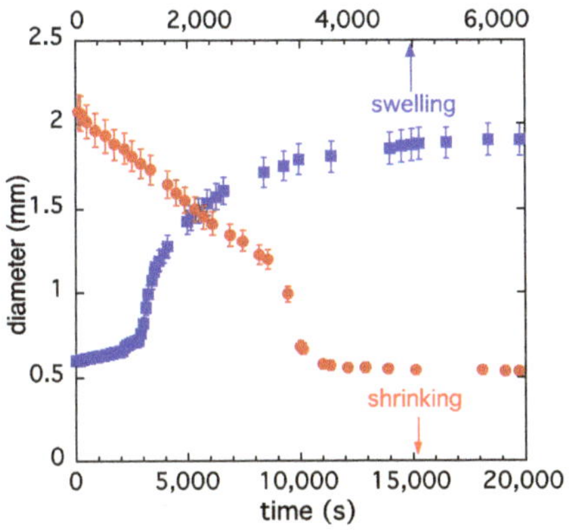

Figure 7. Kinetics of swelling (squares) and shrinking (circles) of a PA hydrogel. Gel was shrunken in 40 mM NaCl solution containing 10 mM $CaCl_2$ solution and reswollen in 40 mM NaCl solution (without $CaCl_2$).

Figure 7 also shows the reswelling of the collapsed PA gel in free Ca^{2+} 40 mM NaCl solution. First, the Ca^{2+} ions diffuse from the collapsed gel into the surrounding NaCl solution, producing a moderate increase in gel swelling. When the Ca^{2+} content of the gel falls below the transition concentration, a rapid increase in the swelling degree takes place, followed by a plateau region, which slightly increases as Ca^{2+} ions gradually leave the gel.

Measurement of the swelling rate makes it possible to estimate the relative 'stability' of the collapsed state of the gels. Figure 8 compares the rate of water uptake of PA gels containing various multivalent cations. To ensure a fair comparison between the swelling rates, prior to the swelling experiment, identical gel samples were collapsed in salt solutions containing 2 mM multivalent salt in 40 mM NaCl. Then, the collapsed gels were transferred into 40 mM NaCl solution. The reswelling of gels collapsed in solutions of alkaline earth metal salts ($CaCl_2$, $SrCl_2$) was the fastest, followed by gels deswollen in solutions of transition metal salts ($CoCl_2$, $NiCl_2$). For gels with trivalent cations (La^{3+}, Ce^{3+}), no appreciable reswelling was observed, even after 3–4 weeks. On the basis of the swelling curves shown in Figure 8, the stability of the collapsed state varies in the order $La^{3+} \approx Ce^{3+} > Ni^{2+} > Co^{2+} > Ca^{2+} \approx Sr^{2+}$. These results are in qualitative agreement with the results of the osmotic and mechanical observations discussed above.

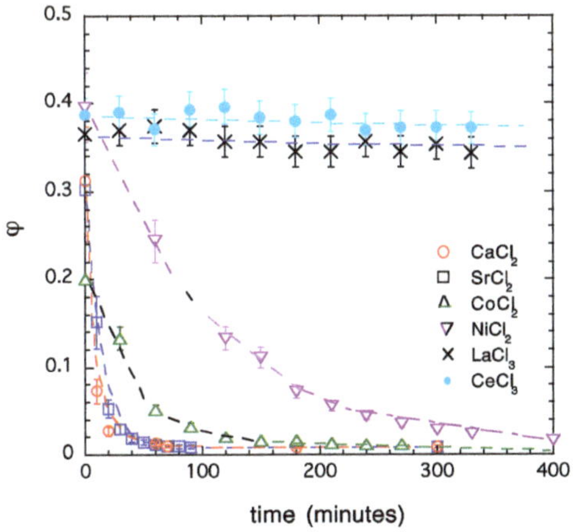

Figure 8. Kinetics of reswelling of PA gels deswollen in multivalent salt solutions. The gels were immersed in 40 mM NaCl solution.

9. Conclusions

The effects of different cations on the osmotic behavior and swelling kinetics of chemically cross-linked PA and DNA gels were discussed. The addition of multivalent cations to polyelectrolyte gels swollen in NaCl solution led to a volume transition in these gels in a biologically relevant concentration range. The electrostatic repulsion between the charged network chains decreased because multivalent ions more efficiently compensate the charge on the polyanion than monovalent counter-ions.

A hierarchy was established in the interaction strength between cations and polyelectrolyte chains according to the chemical group to which the cation belongs. Alkali metal ions (Li^+, Na^+, K^+, Rb^+, Cs^+) practically moved freely all over the entire network. Alkaline earth metal ions (Ca^{2+}, Sr^{2+}) promoted weak associations among the network chains, while transition metal ions (Co^{2+}, Ni^{2+}) formed stronger (but not irreversible) bridges. Trivalent cations (La^{3+} and Ce^{3+}) practically bound irreversibly to the polyanion.

Analysis of the osmotic results on the basis of the Flory–Huggins formalism provided an empirical description of the effect of Ca^{2+} ions on the osmotic pressure both below and in the vicinity of the volume transition. Ca^{2+} ions primarily affected the third-order interaction parameter, which strongly increased with the increasing Ca^{2+} concentration in the surrounding solution, while the second-order interaction parameter only weakly varied. The gradual increase in the interaction parameters with the increasing $CaCl_2$ concentration created the condition for a volume transition. The reversible nature of the volume transition and the absence of a measurable effect of Ca^{2+} ions on the elastic modulus indicated that calcium ion binding is not permanent.

It was shown that the kinetics of the swelling of PA gels in NaCl solution was characterized by a collective diffusion coefficient, which is independent of the initial size of the gel particles. When a gel swollen in NaCl solution is immersed in a solution that contains Ca^{2+} ions, three different stages in the shrinking process can be distinguished. First, the swelling degree slowly decreases with the time. This is followed by a sudden shrinking due to the volume transition. Then, gel contraction continues until the fully collapsed state is reached. When the gel is reswelling from the collapsed state in NaCl solution, first, the swelling degree slowly increases. This is followed by a steep increase corresponding to the volume transition. The last stage is a slow swelling, while the Ca^{2+} concentration gradually decreases in the gel.

In this article, we demonstrated that an ion-induced volume transition in polyelectrolyte gels exhibits a universal behavior, which is practically independent of the molecular details. Although the molecular mechanism responsible for the volume transition is not fully understood, the present results clearly indicate the important role of ion exchange and ion diffusion in the development of complex structures consisting of coexisting shrunken and swollen regions. Understanding the organization of charged macromolecules in near-physiological salt solutions may shed light on the mechanism of structure formation in biological systems, e.g., organization of nucleic acids or other structural elements of the cell or the extracellular matrix. The complexity of the structure and interactions in living systems makes it difficult to perform conclusive experiments under well-controlled conditions. Systematic studies conducted on model gel systems can provide vital insight into the nature of certain universal phenomena that play a role in biological systems. This understanding cannot be obtained from measurements conducted on biological samples because their composition and physical properties cannot be independently and systematically varied as they can be in model systems.

10. Materials and Methods

10.1. Gel Preparation

Sodium polyacrylate gels were produced by free radical copolymerization of partially neutralized acrylic acid and N,N′-methylenebis(acrylamide) cross-linker in aqueous solution according to a procedure described previously [44]. The monomer concentration was 30% (*w/w*), and 35% of the monomers were neutralized by sodium hydroxide before polymerization. Dissolved oxygen was removed by bubbling nitrogen through the solution. Then, ammonium persulfate (0.5 g/L) was added to initiate the polymerization reaction. Gelation was carried out at 80 °C.

Gel beads were produced by polymerization in silicone oil (viscosity: 1000 cPs) that was previously degassed with nitrogen. Spherical droplets (diameter < 1 mm) of the mixture were injected into the silicone oil. After gelation, gel samples were completely neutralized and washed in deionized water to remove all extractable materials (e.g., sol fraction). Water was renewed every day for two weeks.

For the mechanical measurements, cylindrical gel specimens (1 cm height, 1 cm diameter) were produced in a special mold using the same cross-linking procedure. Gel cylinders were neutralized and washed for several weeks with deionized water before the experiments.

DNA gels were produced from deoxyribonucleic acid sodium salt (Na-DNA from salmon testes, Sigma-Aldrich, St. Louis, MO, USA). According to the manufacturer, the % G-C content of this DNA was 41.2%, and the melting temperature was reported to be 87.5 °C in 0.15 M sodium chloride plus 0.015 M sodium citrate. The molecular weight determined by ultracentrifugation was 1.3×10^6, corresponding to approximately 2000 base pairs. First, DNA was dissolved in a HEPES buffer (pH = 7.0); then, the solutions were dialyzed against distilled water. DNA gels were produced by cross-linking [45] with ethylene glycol diglycidyl ether (2%) at pH = 9.0 using TEMED to adjust the pH. The DNA concentration at cross-linking was 3% (*w/w*). The gels were equilibrated in NaCl solutions containing different amounts of $CaCl_2$ (0–0.2 mM).

10.2. Osmotic Stress Measurements

Osmotic stress measurements were conducted on PA and DNA gels by aqueous solutions of poly(vinyl pyrrolidone) (PVP, molecular weight: 29 kDa). The osmotic pressure of the PVP solution was known from independent measurements [54,55]. The swollen network was separated from the solution by a semipermeable membrane, which prevented the penetration of the polymer molecules into the gel. At equilibrium, the swelling pressure of the gel inside the dialysis bag is equal to the osmotic pressure of the PVP solution outside. The size and the weight of the gel samples were measured when equilibrium was attained. The reversibility of the deswelling process was checked by transferring the deswollen gels into PVP solutions of different concentrations. No significant difference was found between swelling degrees obtained by decreasing or increasing the osmotic pressure of the equilibrium solution.

When gel beads were equilibrated with salt solutions, it was assumed that the salt concentration in the liquid phase surrounding the gel sample was unchanged (infinite bath).

10.3. Swelling Kinetics Measurements

Gel beads prepared according to the procedure described above were placed into a Petri dish containing salt solution. The diameter of the gel was measured as a function of time under a Leica MZ 12 stereomicroscope using a calibrated scale. Swelling kinetics measurements were carried out in NaCl solutions containing different amounts of $CaCl_2$ at room temperature [53].

10.4. Elastic Modulus Measurements

Uniaxial compression measurements were performed on gel cylinders in equilibrium with salt solutions using a TA.XT2I HR Texture Analyser (Stable Micro Systems, Vienna Court, UK). This apparatus measures the deformation (±0.001 mm) as a function of the applied force (±0.01 N). Measurements were performed at deformation ratios of $0.7 < \Lambda < 1$. Typical sample sizes were: height 0.5 to 2 cm, diameter 0.5 to 2 cm. The elastic (shear) modulus, G, was calculated from the nominal stress, σ (force per unit undeformed cross-section), using the equation [43]

$$\sigma = G\,(\Lambda - \Lambda^{-2}) \tag{11}$$

where the deformation ratio is $\Lambda = L/L_o$ (L and L_o denote the heights of the deformed and undeformed gel cylinders, respectively).

Both swelling and mechanical measurements were carried out at 25 ± 0.1 °C.

Funding: This research was supported by the Intramural Research Program of the NICHD, NIH.

Data Availability Statement: The data that support the findings of this study are available from the corresponding author upon reasonable request.

Conflicts of Interest: The author declares no conflict of interest.

References

1. Almdal, K.; Dyre, J.; Hvidt, S.; Kramer, O. Towards a Phenomenological Definition of the Term 'Gel'. *Polym. Gels Netw.* **1993**, *1*, 5–17. [CrossRef]
2. Flory, P.J. Constitution of three-dimensional polymers and the theory of gelation. *J. Phys. Chem.* **1942**, *46*, 132–140. [CrossRef]
3. Ferry, J.D. *Viscoelastic Properties of Polymers*, 3rd ed.; John Wiley & Sons: New York, NY, USA, 1980; pp. 529–530.
4. Flory, P.J. *Principles of Polymer Chemistry*; Cornell University: Ithaca, NY, USA, 1953.
5. De Gennes, P.G. *Scaling Concepts in Polymer Physics*; Cornell University Press: Ithaca, NY, USA, 1979.
6. Dusek, K. (Ed.) *Responsive Gels: Volume Transitions*; (Adv. Polym. Sci. Vol. 109); Springer: Berlin/Heidelberg, Germany, 1993.
7. Okano, T. (Ed.) *Biorelated Polymers and Gels—Controlled Release and Applications in Biomedical Engineering*; Academic Press: San Diego, CA, USA, 1998.
8. Siegel, R.A. Stimuli Sensitive Polymers and Self-Regulated Drug Delivery Systems: A Very Partial Review. *J. Control. Release* **2014**, *190*, 337–351. [CrossRef]
9. Roy, I.; Gupta, M.N. Smart Polymeric Materials: Emerging Biochemical Applications. *Chem. Biol.* **2003**, *10*, 1161–1171. [CrossRef] [PubMed]
10. Taylor, M.J.; Chauhan, K.P.; Tarsem, S.; Sahota, T.S. Gels for constant and smart delivery of insulin. *Br. J. Diabetes* **2020**, *20*, 41–51. [CrossRef]
11. Mantha, S.; Pillai, S.; Khayambashi, P.; Upadhyay, A.; Zhang, Y.; Tao, O.; Pham, H.M.; Tran, S.D. Smart Hydrogels in Tissue Engineering and Regenerative Medicine. *Materials* **2019**, *12*, 3323. [CrossRef] [PubMed]
12. Stile, R.A.; Burghardt, W.R.; Healy, K.E. Synthesis and characterization of inject- able poly(N-isopropylacrylamide)-based hydrogels that support tissue formation in vitro. *Macromolecules* **1999**, *32*, 7370–7379. [CrossRef]
13. Treloar, L.R.G. *The Physics of Rubber Elasticity*; Clarendon: Oxford, UK, 1976.
14. Orkoulas, G.; Kumar, S.K.; Panagiotopoulos, A.Z. Monte Carlo study of coulombic criticality in polyelectrolytes. *Phys. Rev. Lett.* **2003**, *90*, 048303. [CrossRef]
15. Yan, Q.; de Pablo, J.J. Hyper-parallel tempering Monte Carlo: Application to the Lennard-Jones fluid and the restricted primitive model. *J. Chem. Phys.* **1999**, *111*, 9509–9516. [CrossRef]
16. Chremos, A.; Douglas, J.F. Polyelectrolyte association and solvation. *J. Chem. Phys.* **2018**, *149*, 163305. [CrossRef]
17. Liao, Q.; Dobrynin, A.V.; Rubinstein, M. Molecular Dynamics Simulations of Polyelectrolyte Solutions: Nonuniform Stretching of Chains and Scaling Behavior. *Macromolecules* **2003**, *36*, 3386–3398. [CrossRef]
18. Carrillo, J.-M.Y.; Dobrynin, A.V. Polyelectrolytes in Salt Solutions: Molecular Dynamics Simulations. *Macromolecules* **2011**, *44*, 5798–5816. [CrossRef]
19. Kenkare, N.R.; Hall, C.K.; Khan, S.A. Theory and simulation of the swelling of polymer gels. *J. Chem. Phys.* **2000**, *113*, 404–418. [CrossRef]
20. Ricka, J.; Tanaka, T. Swelling of Ionic Gels—Quantitative Performance of the Donnan Theory. *Macromolecules* **1984**, *17*, 2916–2921. [CrossRef]
21. Ricka, J.; Tanaka, T. Phase transition in ionic gels induced by copper complexation. *Macromolecules* **1985**, *18*, 83–85. [CrossRef]
22. Kwon, H.J.; Osada, Y.; Gong, J.P. Polyelectrolyte Gels-Fundamentals and Applications. *Polym. J.* **2006**, *38*, 1211–1219. [CrossRef]
23. Hirotsu, S.; Hirokawa, Y.; Tanaka, T. Volume-Phase Transitions of Ionized N-Isopropylacrylamide Gels. *J. Chem. Phys.* **1987**, *87*, 1392–1395. [CrossRef]
24. Bin Imran, A.; Esaki, K.; Gotoh, H.; Seki, T.; Ito, K.; Sakai, Y.; Takeoka, Y. Extremely Stretchable Thermosensitive Hydrogels by Introducing Slide-Ring Polyrotaxane Cross-Linkers and Ionic Groups Into the Polymer Network. *Nat. Commun.* **2014**, *5*, 5124. [CrossRef] [PubMed]
25. Siegel, R.A.; Firestone, B.A. pH-Dependent Equilibrium Swelling Properties of Hydrophobic Poly-Electrolyte Copolymer Gels. *Macromolecules* **1988**, *21*, 3254–3259. [CrossRef]
26. Brannon-Peppas, L.; Peppas, N.A. Equilibrium Swelling Behavior of pH-Sensitive Hydrogels. *Chem. Eng. Sci.* **1991**, *46*, 715–722. [CrossRef]
27. Marcombe, R.; Cai, S.; Hong, W.; Zhao, X.; Lapusta, Y.; Suo, Z. A Theory of Constrained Swelling of a pH-Sensitive Hydrogel. *Soft Matter* **2010**, *6*, 784–793. [CrossRef]
28. Shibayama, M.; Tanaka, T. Volume Phase Transition and Related Phenomena of Polymer Gels. *Adv. Polym. Sci.* **1993**, *109*, 1–62.
29. Katchalsky, A.; Lifson, S.; Eisenberg, H. Equation of Swelling for Polyelectrolyte Gels. *J. Polym. Sci.* **1951**, *7*, 571–574. [CrossRef]

30. Dusek, K.; Patterson, D. Transition in Swollen Polymer Networks Induced by Intramolecular Condensation. *J. Polym. Sci. A* **1968**, *6*, 1209–1216. [CrossRef]
31. Tasaki, I. *Physiology and Electrochemistry of Nerve Fibers*; Academic Press: New York, NY, USA, 1982.
32. DeRossi, D.; Kajiwara, K.; Osada, Y.; Yamauchi, A. *Polymer Gels, Fundamentals and Biomedical Applications*; Plenum Press: New York, NY, USA, 1989.
33. Rubinstein, M.; Papoian, A. Polyelectrolytes in biology and soft matter. *Soft Matter* **2012**, *8*, 9265–9267. [CrossRef]
34. Muthukumar, M. 50th Anniversary Perspective: A Perspective on Polyelectrolyte Solutions. *Macromolecules* **2017**, *50*, 9528–9560. [CrossRef] [PubMed]
35. Dobrynin, A.V.; Colby, R.H.; Rubinstein, M. Scaling Theory of Polyelectrolyte Solutions. *Macromolecules* **1995**, *28*, 1859–1871. [CrossRef]
36. Verdugo, P. Marine Microgels. *Annu. Rev. Mar. Sci.* **2012**, *4*, 375–400. [CrossRef] [PubMed]
37. Verdugo, P.; Alldredge, A.L.; Azam, F.; Kirchman, D.L.; Passow, U.; Santschi, P.H. The oceanic gel phase: A bridge in the DOM-POM continuum. *Mar. Chem.* **2004**, *92*, 67–85. [CrossRef]
38. Verdugo, P.; Orellana, M.V.; Chin, W.C.; Petersen, T.W.; van den Eng, G.; Benner, R.; Hedges, J.I. Marine biopolymer self-assembly: Implications for carbon cycling in the ocean. *Faraday Discuss.* **2008**, *139*, 393–398. [CrossRef] [PubMed]
39. Verdugo, P.; Santschi, P.H. Polymer dynamics of DOC networks and gel formation in seawater. *Deep-Sea Res. Part II* **2010**, *57*, 1489–1493. [CrossRef]
40. Ohmine, I.; Tanaka, T. Salt effects on the phase transition of ionic gels. *J. Chem. Phys.* **1982**, *77*, 5725–5729. [CrossRef]
41. Cheknane, B.; Messaoudene, N.A.; Naceur, M.W.; Zermane, F. Membranes in drinking and industrial water production. *Desalination* **2005**, *179*, 273–280. [CrossRef]
42. Dusek, K.; Prins, W. Structure and elasticity of non-crystalline polymer networks. *Adv. Polym. Sci.* **1969**, *6*, 1–63.
43. Horkay, F.; Tasaki, I.; Basser, P.J. Osmotic swelling of polyacrylate hydrogels in physiological salt solutions. *Biomacromolecules* **2000**, *1*, 84–90. [CrossRef]
44. Horkay, F.; Tasaki, I.; Basser, P.J. Effect of monovalent-divalent cation exchange on the swelling of polyacrylate hydrogels in physiological salt solutions. *Biomacromolecules* **2001**, *2*, 195–199. [CrossRef] [PubMed]
45. Horkay, F.; Basser, P.J. Osmotic Observations on Chemically Cross-Linked DNA Gels in Physiological Salt Solutions. *Biomacromolecules* **2004**, *5*, 232–237. [CrossRef] [PubMed]
46. Kim, B.; Peppas, N.A. Analysis of molecular interactions in poly (methacrylic acid-g-ethylene glycol) hydrogels. *Polymer* **2003**, *44*, 3701–3707. [CrossRef]
47. Peppas, N.A. *Hydrogels in Medicine and Pharmacy*; Chemical Rubber Company: Boca Raton, FL, USA, 1986.
48. Yin, D.W.; Horkay, F.; Douglas, J.F.; de Pablo, J.J. Molecular simulation of the swelling of polyelectrolyte gels by monovalent and divalent counterions. *J. Chem. Phys.* **2008**, *129*, 154902. [CrossRef]
49. Tanaka, T.; Fillmore, D.J. Kinetics of swelling of gels. *J. Chem. Phys.* **1979**, *70*, 1214–1218. [CrossRef]
50. Li, Y.; Tanaka, T. Nonlinear swelling of polymer gels. *J. Chem. Phys.* **1990**, *92*, 1365–1371. [CrossRef]
51. Skuori, R.; Schosseler, F.; Munch, J.P.; Candau, S.J. Swelling and Elastic Properties of Polyelectrolyte Gels. *Macromolecules* **1995**, *28*, 197–210. [CrossRef]
52. Schröder, U.P.; Oppermann, W. Mechanical and stress-optical properties of strongly swollen hydrogels. *Macromol. Chem. Macromol. Symp.* **1993**, *76*, 63–74. [CrossRef]
53. Horkay, F.; Haselkorn, K.; Tasaki, I.; Basser, P.J.; Hecht, A.M.; Geissler, E. Swelling Kinetics of Polyacrylate Gels Beads in Physiological Salt Solutions. *Polym. Prepr.* **2002**, *43*, 573–574.
54. Vink, H. Precision measurements of osmotic pressure in concentrated polymer solutions. *Eur. Polym. J.* **1971**, *7*, 1411–1419. [CrossRef]
55. Horkay, F.; Burchard, W.; Geissler, E.; Hecht, A.M. Thermodynamic Properties of Poly(vinyl alcohol) and Poly(vinyl alcohol-vinyl acetate) Hydrogels. *Macromolecules* **1993**, *26*, 1296–1303. [CrossRef]

gels

MDPI

Review

Dissolved Organic Matter in the Global Ocean: A Primer

Dennis A. Hansell [1,*] and Mónica V. Orellana [2,3]

1 Department of Ocean Sciences, RSMAS, University of Miami, Miami, FL 33149, USA
2 Polar Science Center, Applied Physics Laboratory, University of Washington, Seattle, WA 98105, USA; morellan@uw.edu
3 Institute for Systems Biology, Seattle, WA 98109, USA
* Correspondence: dhansell@rsmas.miami.edu

Abstract: Marine dissolved organic matter (DOM) holds ~660 billion metric tons of carbon, making it one of Earth's major carbon reservoirs that is exchangeable with the atmosphere on annual to millennial time scales. The global ocean scale dynamics of the pool have become better illuminated over the past few decades, and those are very briefly described here. What is still far from understood is the dynamical control on this pool at the molecular level; in the case of this Special Issue, the role of microgels is poorly known. This manuscript provides the global context of a large pool of marine DOM upon which those missing insights can be built.

Keywords: marine dissolved organic carbon; ocean carbon cycle; marine microgels

Citation: Hansell, D.A.; Orellana, M.V. Dissolved Organic Matter in the Global Ocean: A Primer. *Gels* **2021**, *7*, 128. https://doi.org/10.3390/gels7030128

Academic Editor: Maria Valentina Dinu

Received: 28 May 2021
Accepted: 26 August 2021
Published: 28 August 2021

1. Introduction

Dissolved organic carbon (DOC) makes up the second largest bioavailable pools of carbon in the ocean (~660 Pg C (1 Pg = 1 × 10^9 metric tons); [1]) and is second only to the ~50× larger pool of dissolved inorganic carbon. The size of the reservoir, and its complementary functions as a sink for autotrophically fixed carbon and as a source of substrate to microbial heterotrophs, indicate that DOC plays a central role in the ocean carbon cycle [2]. Identifying the details and mechanisms of that role in the global ocean remains a great challenge; with relevance to this Special Issue, the contribution of marine gels to those mechanisms is essentially unknown. While a critical percentage of DOC (varying from 10% in the coastal ocean [3] to 30% in the Arctic [4]) assemble as microgels (Orellana and Hansell, this issue), their importance in DOC basin-scale dynamics is just beginning to be understood [5]. Marine gels are three-dimensional (3D), colloidal- to micrometer-sized hydrogel networks held together by Ca^{2+} ionic bonds/hydrophobic bonds that assemble spontaneously from marine dissolved biopolymers [3–5]; they have been proposed to play a pivotal role in regulating ocean basin-scale biogeochemical dynamics [6].

In this paper, the role of DOC in the ocean carbon cycle is considered in its broadest temporal and spatial scales, largely as a primer for those wishing to understand the global scale dynamics of the pool. The paper begins with an evaluation of the spatial distribution of DOC at the regional and basin scales, in both the surface and deep ocean. In this context, the net production of DOC relative to the distribution and timing of marine primary production is evaluated. It briefly concludes with priorities for present and future research relevant to the role of gels in DOC dynamics. More complete and detailed reviews of DOC dynamics in the ocean are available [7–13].

2. DOC Concentrations and Reactivity

DOC concentrations in the ocean (Figure 1) range from a deep ocean low of ~35 μmolC/kg (in deep waters of the Pacific that were last exposed to the atmosphere and sunlight more than a millennium previously) to surface ocean highs of >80 μmolC/kg, with the highest concentrations commonly found in river-influenced coastal waters [1]. Biological processes establish the vertical gradient seen in Figure 1, with net autotrophic production in the sunlit surface layer and net heterotrophic consumption at depth, while physical conditions maintain the gradient (i.e., high vertical stability in the ocean water column largely precludes ready mixing of DOC-enriched surface waters to great depths).

Bulk DOC in the ocean is operationally resolved into at least three fractions, each qualitatively characterized by its biological lability [8]. All ocean depths contain (1) the very old, biologically refractory DOC (RDOC concentrations <45 μM and bulk radiocarbon ages of >6000 years) [14]. The RDOC distribution is controlled by the global deep overturning circulation and is thus relatively homogeneous in the deep ocean (Figure 1, note blues and pinks). Built upon the refractory DOC, at intermediate (to 1000 m) and upper layer depths, is (2) material of intermediate (or semi-) lability (SLDOC; lifetime of months to years; note greens, yellows, and reds in Figure 1). It is this material that has recently (years) accumulated in the surface ocean and that is then mixed downward into the ocean interior, thereby reducing the vertical concentration gradient and contributing to carbon export (i.e., the biological carbon pump; note great depth of green colors in the far left (near Iceland in the North Atlantic) of Figure 1). Concentrations of this fraction are commonly 10–30 μmolC/kg in the stratified upper ocean, and near zero in the deep ocean, indicating that it is susceptible to removal over decades. The most biologically labile fraction of DOC (3), with lifetimes of days to months and concentrations of just a few to 10′s of μmolC/kg, is largely limited to the sunlit layer of the ocean, where it is produced by autotrophs daily [2]. Referred to as labile DOC (LDOC), this material supports microbial heterotrophic processes in the surface ocean. Of the three fractions, it shows the greatest seasonality, with high net production rates during phytoplankton blooms and lower rates during the autumn and winter convective overturns of the upper ocean.

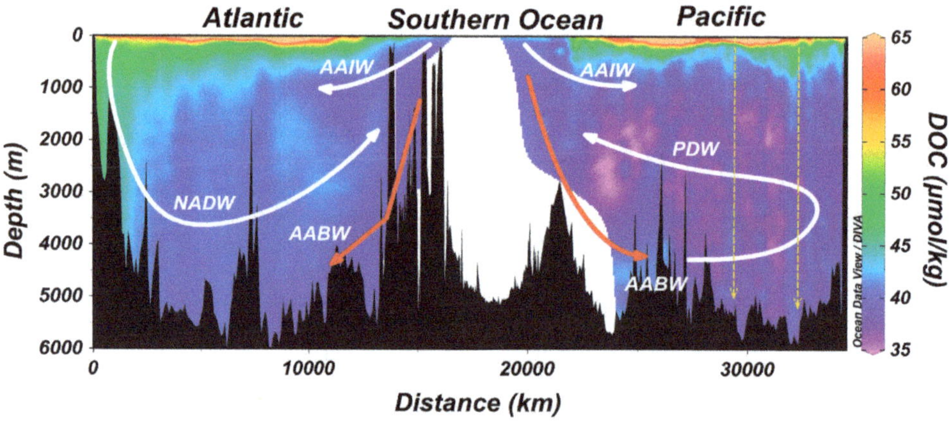

Figure 1. Vertical distribution of DOC (μmolC/kg) along ocean sections A16 and P16 in the Atlantic and Pacific Oceans, respectively (locations given in Figure 2). Data density is low in the Southern Ocean along the section connecting the southern termini of A16 and P16, hence the gap in coverage. Solid arrows schematically indicate the major overturning circulation pathways in the ocean, specifically NADW, AABW, AAIW, and PDW. The dashed arrows represent the apparent pathways of sinking biogenic particles (exported from the surface ocean) that disaggregate while sinking, in turn adding DOC to the deep-water column. This enrichment serves as a substrate to heterotrophs (including prokaryotes) living in the greatest depths of the ocean. Note the vertical alignments of DOC enrichment adjacent to those arrows due to the disaggregation and solubilization of particles. NADW = North Atlantic Deep Water; AABW = Antarctic Bottom Water; AAIW = Antarctic Intermediate Water; PDW = Pacific Deep Water.

3. Global Distribution

There are two important perspectives taken when considering the distribution of DOC in the global ocean [1]. The first is in the wind driven upper layers; the wind forcing of the water column is at its strongest up to ~200 m depth, but it is still significant to depths up of 1000 m in some locations. The second is at greater depths (the deep ocean). The surface ocean is where most of the newly produced DOC that escapes fast microbial consumption (i.e., LDOC) accumulates (i.e., SLDOC). This material can then be delivered to great ocean depths as it is carried downward with the ocean's overturning water masses; these waters ventilate the deep ocean over decadal to millennial time scales, carrying DOC with them (see section below).

The distribution of DOC in the upper ocean is given in Figure 2. The highest concentrations are at low to mid-latitudes, while the lowest concentrations are common at higher latitudes, such as in the Southern Ocean (the Arctic Ocean is a strong exception to this rule, as it receives high loads of DOC created on land and that are delivered via the large Arctic rivers). Surface ocean waters at low to mid-latitudes are resistant to deep vertical mixing because the surface layer density is low due to solar warming, while the deeper underlying waters have high density due to very low temperatures (<3 °C, having originated in polar domains). Consequently, materials (e.g., DOC, heat, salt, planktonic organisms) present in those surface waters tend to remain there; they are not easily moved downward by vertical mixing. At higher latitudes, vertical stability is weaker (water density is high throughout the water column), so mixing is deeper, and the DOC produced at the surface can see its concentration reduced by that mixing.

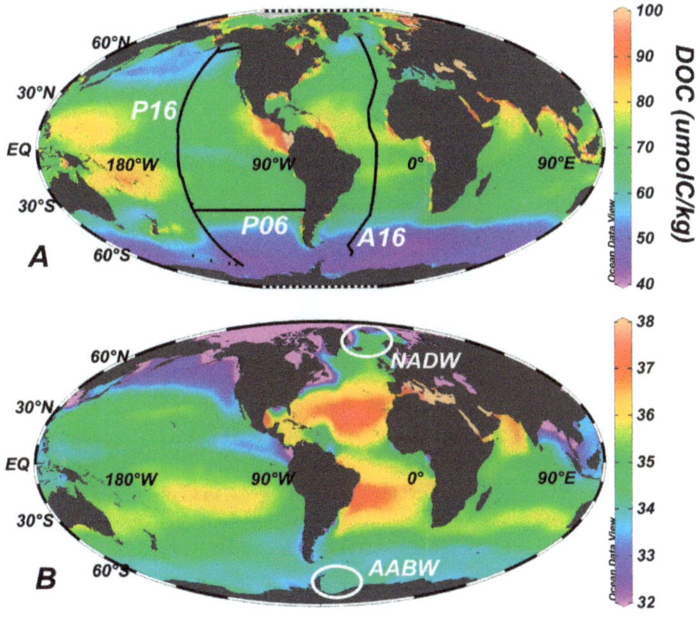

Figure 2. Modeled surface ocean distribution of (**A**) DOC (μmolC/kg) [15] and (**B**) gridded salinity (unitless) in August [16]. Ocean sections A16, P16, and P06 are also shown, the data for which are shown in Figures 1 and 3. Figures were created using Ocean Data View [17]. White ellipses indicate formation locations of North Atlantic Deep Water (NADW) and Antarctic Bottom Water (AABW).

4. Zones of Net DOC Production

It is important to note that concentrations of DOC in the surface ocean are not indicators of net DOC production rates in those locations. Instead, the rates of production are typically low in the most stratified waters (where DOC is elevated) because of the generally low nutrient concentrations there (i.e., oligotrophic conditions). It is largely the high vertical stratification existing in those nutrient impoverished waters that allows

DOC to accumulate to elevated concentrations. The ocean systems producing DOC at the highest rates are typically found where the net productivity by autotrophs is high (i.e., eutrophic upwelling ocean systems) [18]. However, concentrations of DOC in upwelling systems are not typically high because the upwelled waters start with low initial DOC concentrations. However, the change in concentrations because of the upwelling can be large. In Figure 3, we see the vertical distributions of DOC (Figure 3A) and the nutrient phosphate (Figure 3B) off the coast of Chile in the South Pacific. Upwelling is evident by the uplift of high nutrient waters at the coast (note arrow in figure). Phosphate concentrations decrease upon reaching the euphotic zone due to net community production (NCP) in the system. One of the products of marine NCP is DOC; it increases in concentration (in the vertical) at the coast by ~10–15 µmolC/kg (from ~200 m to the surface). This concentration increase, as a product of NCP, is much larger than would be anticipated in the oligotrophic subtropical gyres. Net DOC production, as a function of NCP, is predictable [18–20].

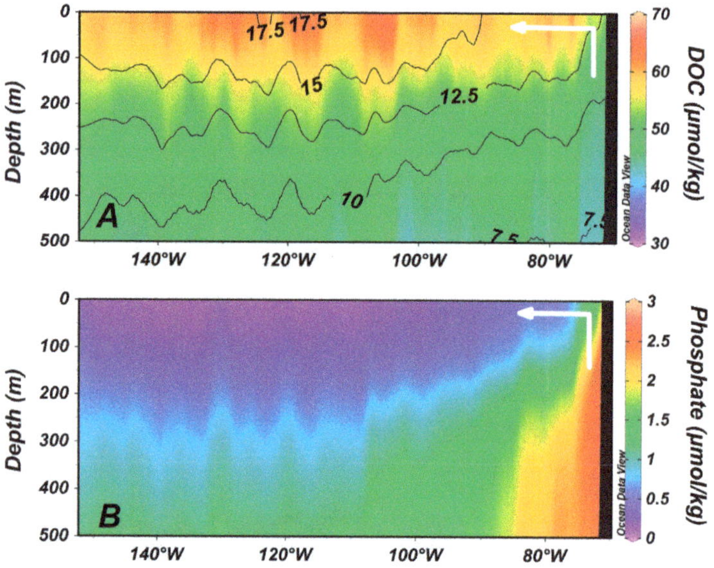

Figure 3. Vertical sections of (A) DOC (µmolC/kg) overlain by temperature contours (°C) and (B) phosphate (µmolP/kg) along ocean section P06 (see Figure 2 for location). White arrows schematically indicate upwelling along the coast of Chile.

5. Export of Surface Accumulated DOC to Depth with the Ocean's Overturning Circulation

DOC accumulated in the upper ocean is susceptible to export down into the ocean interior by overturning circulation [1,21–23]. Such export occurs if the waters enriched in DOC reach the ocean zones that are seasonally made dense enough (through cooling or salinification) to be overturned, thus contributing to surface DOC export to the intermediate/deep ocean. To highlight the process, two ocean regions that are important for deep water formation are shown in Figure 2B. NADW forms north of Iceland, while AABW forms in the Weddell Sea (among other locations around Antarctica). In Figure 1 (far left in plot), we see that DOC-enriched waters are present (green color) throughout the water column; this distribution indicates that DOC was mixed downward during overturn associated with NADW formation, thus contributing to carbon export. NADW formation, then, is an important process for the export of DOC to great ocean depths. We do not see similar enrichments at depth in the Southern Ocean (center of plot in Figure 1) [24] despite the source of AABW being nearby. The difference between these two systems is that the Polar

Frontal Zone of the Southern Ocean prevents DOC-enriched lower latitude surface waters from reaching the areas of bottom water formation. Evidence for the strong boundary created by the frontal system is the gradient in salinity (Figure 2B) from the subtropical gyres (green through red colors) to the Southern Ocean (blue colors). In contrast, the North Atlantic does not hold such a transport boundary; high-salinity and high-DOC lower latitude waters (such as from the subtropical gyre of the North Atlantic) are transported to the deep water formation region, thus favoring DOC export there [25]. There are several other deep and intermediate water formation sites in the ocean [26]; the role of each in DOC export similarly depends on the amount of DOC present in the surface waters at the initiation of the overturn.

Once the DOC-enriched deep waters (e.g., NADW) have formed, the excess DOC is slowly removed with time [20]. The longer the time since export, the less observable a decrease in concentration, though composition does show modest changes [27–29]. Whether the removal processes are biotic or abiotic remains to be demonstrated, as does the function of gels in deep DOC removal. Other papers consider DOC export by mechanisms such as convective overturn and release from sinking biogenic particles [23,30].

6. Zones of Deep Ocean DOC Enrichment Due to Sinking Biogenic Particles

In the section above, enhanced DOC concentrations in the deep ocean, such as in the far North Atlantic, were described as being due to introduction with the overturn of the water column. Biogenic organic particles sinking from the surface ocean likewise will introduce DOC to great depths [31–34]. Evidence for input by this mechanism is seen in Figure 1 (note dashed vertical arrows in the deep Pacific). The arrows are placed over small DOC enrichments (relative to surrounding waters) aligned vertically in the water column. As the particles sink, various biological and abiotic processes lead to their disaggregation and solubilization, with some of that material appearing as DOC [35]. This newly added DOC apparently has a limited lifetime (months) [34] because it serves as a substrate to the deep heterotrophic microbe populations [36]. We expect to find such enrichments spatially distributed throughout the global ocean and that are especially present where the export of large dense particles occurs, as these particles are more likely to reach the greatest depths of the ocean quickly. More slowly sinking particles typically do not reach great depths because they are intercepted by consumers or are disaggregated in the upper layers; hence, they will add DOC to those upper layers but to not the deep layers.

7. Composition of Ocean Dissolved Organic Matter

Dissolved organic matter (DOM) in the ocean can be categorized into two fractions based on radiocarbon ages and molecular composition. The first fraction is freshly produced, phytoplankton-derived DOM that is largely composed of classical biomolecules of known building blocks, such as polysaccharides, proteins, and lipids [37]. This is the material that is initially released by autotrophs in the euphotic zone, leading to accumulation there, and it is the material released into the deeper water column by sinking particles. Given its recent production, its radiocarbon age is modern. This modern material, which holds the LDOC and SLDOC fractions (described in Section 2), constitutes perhaps 3% of the oceanic DOC pool, with most being in the upper ocean layers.

The balance of oceanic DOC (97%) has a much greater radiocarbon age (>4000 years) and is absent in the classical biochemical character of the modern fraction. Most of the dissolved organic compounds in this older fraction have low molecular mass (<1000 Da) [38], and the chemical diversity is analytically challenging to characterize; the molecular structure of only a minor fraction of all of the compounds present is known. Estimates on the number of different compounds in DOM are inexact, but more than 20,000 molecular formulae have been identified with ultrahigh-resolution mass spectrometry, with 30 or more isomers per formula [39]. There are likely millions of different compound structures in DOM, each presumably below pico-molar concentrations. More detailed considerations of DOM composition are available in [37,40–42].

8. Closing with Consideration of Marine Gels

Microgels may account for an important fraction of DOC, having essential roles in shunting DOC polymers into particulate organic carbon (POC) through spontaneous assembly as well as providing polymer gel-rich substrates and habitats for bacterial biodegradation and remineralization. Gels have been proposed to play a pivotal role in regulating ocean basin-scale biogeochemical dynamics [6]. Marine gels link biological production at the ocean's surface and microbial degradative processes at the ocean's interior, cloud properties, radiative balance, and global climate [6].

Microgels exhibit unique physicochemical characteristics, such as reversible volume phase transitions stimulated by diverse environmental forcing conditions, such as pH [3,43], temperature [44,45], DMS and DMSP concentrations [4], solvent composition, light [46], etc. Though we still do not know the quantitative role of marine gels in the ocean carbon cycle, environmental effects on DOC biopolymers might have serious consequences for DOC dynamics and bioavailability. However, there are many issues that we still need to consider, such as what are the kinetics of polymer assembly, in situ? How does pressure affect assembly? What are the microgel distributions in the water column in the different oceans? What is their variability in the coastal ocean versus the open ocean? What are the ages of the polymer gels? What are the compositions of the assembled polymers, and how does it vary with depth? Answering these questions and many more will allow us to accurately understand the mechanisms involved in the transformation and dynamics of DOC polymers, and most importantly allow us to predict the effects of climate/ocean change and acidification on DOC dynamics. Marine polymer dynamics viewed in the context of soft matter physics can provide clear benefits for developing accurate models of the response of biogeochemical cycles to environmental forcing.

Author Contributions: Conceptualization, D.A.H. and M.V.O.; writing—original draft preparation, D.A.H.; writing—review and editing, M.V.O. Both authors have read and agreed to the published version of the manuscript.

Funding: This work was funded by the U.S. NSF OCE-1436748 and OCE-2023500 and U.S. NASA 80NSSC18K0437 to DAH and U.S. NSF OCE-1634009 to MVO.

Data Availability Statement: Data plotted in the Figures 1 and 3 are available at the NOAA National Centers for Environmental Information [47]. Data sources for Figure 2 are given in the caption.

Conflicts of Interest: The authors declare no conflict of interest. The funders had no role in the design of the study; in the collection, analyses, or interpretation of data; in the writing of the manuscript; or in the decision to publish the results.

References

1. Hansell, D.A.; Carlson, C.A.; Repeta, D.J.; Schlitzer, R. Dissolved organic matter in the ocean: A controversy stimulates new insights. *Oceanography* **2009**, *22*, 202–211. [CrossRef]
2. Carlson, C.A.; Hansell, D.A. DOC sources, sinks and fluxes. In *Biogeochemistry of Marine Dissolved Organic Matter*, 2nd ed.; Elsevier: London, UK, 2014; pp. 65–126.
3. Chin, W.C.; Orellana, M.V.; Verdugo, P. Spontaneous assembly of marine dissolved organic matter into polymer gels. *Nature* **1998**, *391*, 568–572. [CrossRef]
4. Orellana, M.V.; Matrai, P.A.; Leck, C.; Rauschenberg, C.D.; Lee, A.M.; Coz, E. Marine microgels as a source of cloud condensation nuclei in the high Arctic. *Proc. Natl. Acad. Sci. USA* **2011**, *108*, 13612–13617. [CrossRef]
5. Verdugo, P. Marine Microgels. In *Annual Review of Marine Science*; Annual Reviews: San Mateo, CA, USA, 2012; Volume 4, pp. 375–400.
6. Orellana, M.V.; Leck, C. Marine Microgels. In *Biogeochemistry of Marine Dissolved Organic Matter*, 2nd ed.; Academic Press: Boston, MA, USA, 2015; pp. 451–480. [CrossRef]
7. Hansell, D.A.; Carlson, C.A. *Biogeochemistry of Marine Dissolved Organic Matter*, 2nd ed.; Elsevier: Waltham, MA, USA, 2014.
8. Hansell, D.A. Recalcitrant Dissolved Organic Carbon Fractions. In *Annual Review of Marine Science*; Annual Reviews: San Mateo, CA, USA, 2013; Volume 5, pp. 421–445.
9. Lonborg, C.; Carreira, C.; Jickells, T.; Alvarez-Salgado, X.A. Impacts of global change on ocean dissolved organic carbon (DOC) cycling. *Front. Mar. Sci.* **2020**, *7*, 466. [CrossRef]

10. Bauer, J.E.; Bianchi, T.S. Dissolved Organic Carbon Cycling and Transformation. In *Treatise on Estuarine and Coastal Science*; Wolanski, E., McLusky, D.S., Eds.; Academic Press: Waltham, MA, USA, 2011; Volume 5, pp. 7–67.
11. Ogawa, H.; Tanoue, E. Dissolved organic matter in oceanic waters. *J. Oceanogr.* **2003**, *59*, 129–147. [CrossRef]
12. Moran, M.A.; Kujawinski, E.B.; Stubbins, A.; Fatland, R.; Aluwihare, L.I.; Buchan, A.; Crump, B.C.; Dorrestein, P.C.; Dyhrman, S.T.; Hess, N.J.; et al. Deciphering ocean carbon in a changing world. *Proc. Natl. Acad. Sci. USA* **2016**, *113*, 3143–3151. [CrossRef] [PubMed]
13. Wagner, S.; Schubotz, F.; Kaiser, K.; Hallmann, C.; Waska, H.; Rossel, P.E.; Hansmann, R.; Elvert, M.; Middelburg, J.J.; Engel, A.; et al. Soothsaying DOM: A current perspective on the future of oceanic dissolved organic carbon. *Front. Mar. Sci.* **2020**, *7*, 341. [CrossRef]
14. Bauer, J.E.; Williams, P.M.; Druffel, E.R.M. C-14 Activity of dissolved organic carbon fractions in the North-central Pacific and Sargasso Sea. *Nature* **1992**, *357*, 667–670. [CrossRef]
15. Roshan, S.; DeVries, T. Efficient dissolved organic carbon production and export in the oligotrophic ocean. *Nat. Commun.* **2017**, *8*, 8. [CrossRef] [PubMed]
16. Schlitzer, R. *Ocean Data View*; Alfred Wegener Institute: Bremerhaven, Germany, 2021.
17. Hansell, D.A.; Carlson, C.A.; Amon, R.M.W.; Álvarez-Salgado, X.A.; Yamashita, Y.; Romera-Castillo, C.; Bif, M.B. *Compilation of Dissolved Organic Matter (DOM) Data Obtained from Global Ocean Observations from 1994 to 2020*; NOAA National Centers for Environmental Information: Washington, DC, USA, 2021. [CrossRef]
18. Hansell, D.A.; Carlson, C.A. Net community production of dissolved organic carbon. *Glob. Biogeochem. Cycles* **1998**, *12*, 443–453. [CrossRef]
19. Bif, M.B.; Hansell, D.A. Seasonality of dissolved organic carbon in the upper northeast Pacific Ocean. *Glob. Biogeochem. Cycles* **2019**, *33*, 526–539. [CrossRef]
20. Romera-Castillo, C.; Letscher, R.T.; Hansell, D.A. New nutrients exert fundamental control on dissolved organic carbon accumulation in the surface Atlantic Ocean. *Proc. Natl. Acad. Sci. USA* **2016**, *113*, 10497–10502. [CrossRef]
21. Carlson, C.A.; Hansell, D.A.; Nelson, N.B.; Siegel, D.A.; Smethie, W.M.; Khatiwala, S.; Meyers, M.M.; Halewood, E. Dissolved organic carbon export and subsequent remineralization in the mesopelagic and bathypelagic realms of the North Atlantic basin. *Deep-Sea Res. Part Ii-Top. Stud. Oceanogr.* **2010**, *57*, 1433–1445. [CrossRef]
22. Hansell, D.A.; Carlson, C.A.; Schlitzer, R. Net removal of major marine dissolved organic carbon fractions in the subsurface ocean. *Glob. Biogeochem. Cycles* **2012**, *26*, GB1016. [CrossRef]
23. Baetge, N.; Graff, J.R.; Behrenfeld, M.J.; Carlson, C.A. Net community production, dissolved organic carbon accumulation, and vertical export in the western North Atlantic. *Front. Mar. Sci.* **2020**, *7*, 227. [CrossRef]
24. Bercovici, S.K.; Huber, B.A.; DeJong, H.B.; Dunbar, R.B.; Hansell, D.A. Dissolved organic carbon in the Ross Sea: Deep enrichment and export. *Limnol. Oceanogr.* **2017**, *62*, 2593–2603. [CrossRef]
25. Fontela, M.; Garcia-Ibanez, M.I.; Hansell, D.A.; Mercier, H.; Perez, F.F. Dissolved Organic Carbon in the North Atlantic Meridional Overturning Circulation. *Sci. Rep.* **2016**, *6*, 26931. [CrossRef]
26. Hanawa, K.; Talley, L.D. Chapter 5.4 Mode waters. In *International Geophysics*; Siedler, G., Church, J., Gould, J., Eds.; Academic Press: Cambridge, MA, USA, 2001; Volume 77, pp. 373–386.
27. Broek, T.A.B.; Walker, B.D.; Guilderson, T.P.; Vaughn, J.S.; Mason, H.E.; McCarthy, M.D. Low molecular weight dissolved organic carbon: Aging, compositional changes, and selective utilization during global ocean circulation. *Glob. Biogeochem. Cycles* **2020**, *34*, e2020GB006547. [CrossRef]
28. Flerus, R.; Lechtenfeld, O.J.; Koch, B.P.; McCallister, S.L.; Schmitt-Kopplin, P.; Benner, R.; Kaiser, K.; Kattner, G. A molecular perspective on the ageing of marine dissolved organic matter. *Biogeosciences* **2012**, *9*, 1935–1955. [CrossRef]
29. Bercovici, S.K.; Koch, B.P.; Lechtenfeld, O.J.; McCallister, S.L.; Schmitt-Kopplin, P.; Hansell, D.A. Aging and molecular changes of dissolved organic matter between two deep oceanic end-members. *Glob. Biogeochem. Cycles* **2018**, *32*, 1449–1456. [CrossRef]
30. Turner, J.T. Zooplankton fecal pellets, marine snow, phytodetritus and the ocean's biological pump. *Prog. Oceanogr.* **2015**, *130*, 205–248. [CrossRef]
31. Noji, T.T.; Borsheim, K.Y.; Rey, F.; Nortvedt, R. Dissolved organic carbon associated with sinking particles can be crucial for estimates of vertical carbon flux. *Sarsia* **1999**, *84*, 129–135. [CrossRef]
32. Orellana, M.V.; Hansell, D.A. Ribulose-1,5-bisphosphate carboxylase/oxygenase (RuBisCO): A long-lived protein in the deep ocean. *Limnol. Oceanogr.* **2012**, *57*, 826–834. [CrossRef]
33. Collins, J.R.; Edwards, B.R.; Thamatrakoln, K.; Ossolinski, J.E.; DiTullio, G.R.; Bidle, K.D.; Doney, S.C.; Van Mooy, B.A.S. The multiple fates of sinking particles in the North Atlantic Ocean. *Glob. Biogeochem. Cycles* **2015**, *29*, 1471–1494. [CrossRef]
34. Lopez, C.N.; Robert, M.; Galbraith, M.; Bercovici, S.K.; Orellana, M.V.; Hansell, D.A. High temporal variability of total organic carbon in the deep northeastern Pacific. *Front. Earth Sci.* **2020**, *8*, 80. [CrossRef]
35. Repeta, D.J.; Aluwihare, L.I. Radiocarbon analysis of neutral sugars in high-molecular-weight dissolved organic carbon: Implications for organic carbon cycling. *Limnol. Oceanogr.* **2006**, *51*, 1045–1053. [CrossRef]
36. Hansman, R.L.; Griffin, S.; Watson, J.T.; Druffel, E.R.M.; Ingalls, A.E.; Pearson, A.; Aluwihare, L.I. The radiocarbon signature of microorganisms in the mesopelagic ocean. *Proc. Natl. Acad. Sci. USA* **2009**, *106*, 6513–6518. [CrossRef] [PubMed]
37. Repeta, D.J. Chemical Characterization and Cycling of Dissolved Organic Matter. In *Biogeochemistry of Marine Dissolved Organic Matter*, 2nd ed.; Elsevier: London, UK, 2014; pp. 21–63.

38. Riedel, T.; Dittmar, T. A method detection limit for the analysis of natural organic matter via Fourier Transform Ion Cyclotron Resonance Mass Spectrometry. *Anal. Chem.* **2014**, *86*, 8376–8382. [CrossRef]
39. Hertkorn, N.; Ruecker, C.; Meringer, M.; Gugisch, R.; Frommberger, M.; Perdue, E.M.; Witt, M.; Schmitt-Kopplin, P. High-precision frequency measurements: Indispensable tools at the core of the molecular-level analysis of complex systems. *Anal. Bioanal. Chem.* **2007**, *389*, 1311–1327. [CrossRef]
40. Nebbioso, A.; Piccolo, A. Molecular characterization of dissolved organic matter (DOM): A critical review. *Anal. Bioanal. Chem.* **2013**, *405*, 109–124. [CrossRef] [PubMed]
41. Baltar, F.; Alvarez-Salgado, X.A.; Aristegui, J.; Benner, R.; Hansell, D.A.; Herndl, G.J.; Lonborg, C. What Is refractory organic matter in the ocean? *Front. Mar. Sci.* **2021**, *8*, 327. [CrossRef]
42. Arakawa, N.; Aluwihare, L.I.; Simpson, A.J.; Soong, R.; Stephens, B.M.; Lane-Coplen, D. Carotenoids are the likely precursor of a significant fraction of marine dissolved organic matter. *Sci. Adv.* **2017**, *3*, 11. [CrossRef] [PubMed]
43. Tanaka, T.; Fillmore, D.; Sun, S.-T.; Nishio, I.; Swislow, G.; Shah, A. Phase transitions in ionic gels. *Phys. Rev. Lett.* **1980**, *45*, 1636–1639. [CrossRef]
44. Shibayama, M.; Tanaka, T.; Han, C.C. Small angle neutron scattering study on poly(N-isopropyl acrylamide) gels near their volume-phase transition temperature. *J. Chem. Phys.* **1992**, *97*, 13. [CrossRef]
45. Chen, C.-S.; Anaya, J.M.; Chen, E.Y.T.; Farr, E.; Chin, W.-C. Ocean Warming–Acidification Synergism Undermines Dissolved Organic Matter Assembly. *PLoS ONE* **2015**, *10*, e0118300. [CrossRef] [PubMed]
46. Tanaka, T. Phase transitions of gels. In *Polyelectrolyte Gels*; American Chemical Society: Washington, DC, USA, 1992; Volume 480, pp. 1–21.
47. Boyer, T.P.; Antonov, J.I.; Baranova, O.K.; Coleman, C.; Garcia, H.E.; Grodsky, A.; Johnson, D.R.; Locarnini, R.A.; Mishonov, A.V.; O'Brien, T.D.; et al. *World Ocean Database 2013*; National Oceanic and Atmospheric Administration: Washington, DC, USA, 2013. [CrossRef]

gels

MDPI

Review

Marine Biopolymer Dynamics, Gel Formation, and Carbon Cycling in the Ocean

Pedro Verdugo

Department of Bioengineering, University of Washington, Friday Harbor Laboratories, Friday Harbor, WA 98250, USA; verdugo@uw.edu

Abstract: Much like our own body, our planet is a macroscale dynamic system equipped with a complex set of compartmentalized controls that have made life and evolution possible on earth. Many of these global autoregulatory functions take place in the ocean; paramount among those is its role in global carbon cycling. Understanding the dynamics of organic carbon transport in the ocean remains among the most critical, urgent, and least acknowledged challenges to modern society. Dissolved in seawater is one of the earth's largest reservoirs of reduced organic carbon, reaching ~700 billion tons. It is composed of a polydisperse collection of marine biopolymers (MBP), that remain in reversible assembled↔dissolved equilibrium forming hydrated networks of marine gels (MG). MGs are among the least understood aspects of marine carbon dynamics. Despite the polymer nature of this gigantic pool of material, polymer physics theory has only recently been applied to study MBP dynamics and gel formation in the ocean. There is a great deal of descriptive phenomenology, rich in classifications, and significant correlations. Still missing, however, is the guide of robust physical theory to figure out the fundamental nature of the supramolecular interactions taking place in seawater that turn out to be critical to understanding carbon transport in the ocean.

Keywords: dissolved organic matter; biopolymer self-assembly; marine gels; phytoplankton exocytosis; volume phase transition; bacterial colonization; polymer networks theory; reactive organic matter; recalcitrant organic matter; global carbon cycling

Citation: Verdugo, P. Marine Biopolymer Dynamics, Gel Formation, and Carbon Cycling in the Ocean. *Gels* **2021**, *7*, 136. https://doi.org/10.3390/gels7030136

Academic Editor: Jean-Michel Guenet

Received: 8 July 2021
Accepted: 1 September 2021
Published: 9 September 2021

1. Introduction

By the end of the nineties, I made acquaintance with John Hedges, not because of the ocean, but because we both were farm boys: he took care of pigs, I did it with cattle; there was a lot of smell to share. In the winter of 2000, John sent me a set of samples of freshly filtered seawater (SW) to find out the size distribution of the molecules present in these samples. As expected, the laser spectrometer detected a broad polydispersity dominated by small nanometer-size species. Inadvertently, one sample was left in the spectrometer, and a few days later Prof. Wei Chun Chin—a graduate student at that time—reported that now the spectrometer was detecting particles of several microns size. The initial thought was that bacterial contamination might be taking place and was probably forming colonies. However, unlike particles that typically undergo continuous random walk with a characteristic Gaussian profile, bacteria—as discovered by my friend Ralph Nossal—move in a random go-stop Markovian fashion with a spectral signature of a Poisson profile [1]. The particles were not bacteria. Further studies indicated that these particles were microscopic gels. It opens the trail to the first objective demonstration [2] that marine biopolymer (MBP) dissolved in seawater (SW) could undergo self-assembly forming marine gels (MG). The broad significance of these observations [3] prompted us to cast a wide effort to apply theory and methods of polymer physics to test specific hypotheses to investigate carbon dynamics in the ocean. Engineering at NSF generously funded the idea. Unfortunately, however, by the time funding arrived, John had unexpectedly left us. This paper is dedicated to his memory.

Avoiding unnecessary formalization, here is a perspective of MBP dynamics in the light of physical theory. Part of it has been thoroughly verified in the laboratory and published; other, reported at meetings—although compelling and supported by robust data—are outcomes resulting from limited field sampling that requires thorough verification. They are included as inviting head trails to open further exploration.

The focus, illustrated in Figure 1, is constrained to studies on how (1) MBP are born, (2) self-assembled in SW, (3) presented as a concentrated MG substrate to microorganisms, and (4) finally cleaved and discarded as refractory stock in the ocean. To avoid confusion, biopolymers found in SW, which are a complex pool of different origins and are certainly not transparent, will be simply labeled as marine biopolymers (MBP). Similarly, marine gels (MG) are labeled as such, not as transparent exopolymer particles (TEP) since they are made of polymers of many sources not only exopolymers.

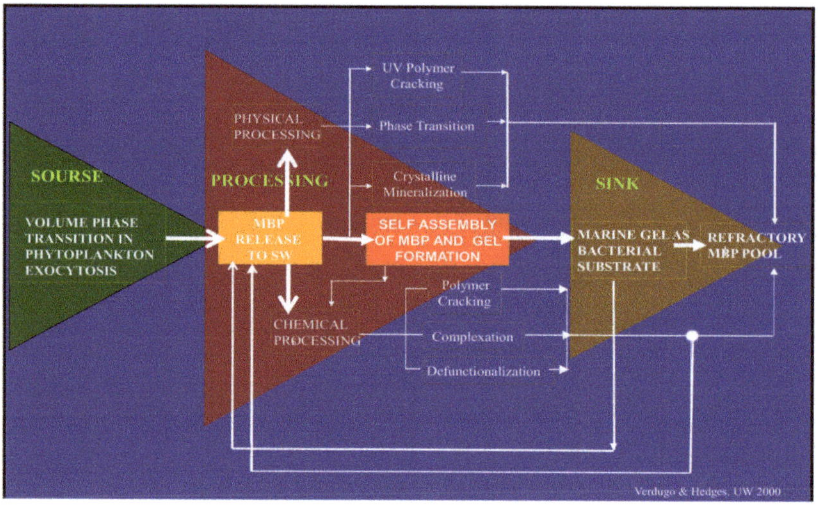

Figure 1. Before their release to SW, MBP are held inside phytoplankton secretory granules. They are gels stored in a condensed phase inside the cell. Upon export from the cell by exocytosis they undergo typical polymer gel phase transition to solvated phase with massive swelling. They are finally released as loosely woven physical hydrogels. Once discharged to SW, MBP can separate, disperse broadly to join the dissolved organic matter (DOM) pool. While in the DOM pool, MBP can undergo multiple physical or chemical processes. Among those, self-assembly is the one that results in the formation of MGs. Marine gels contain 10^4 more organic material than their MBP dissolved precursors. MG form discrete patches of concentrated substrate that bacterial enzymes can readily cleave [4]. While cleaved residues <600 D can be readily incorporated and metabolized by marine bacteria (MB) and thereby enter the food chain to higher trophic levels, larger cleaved residues can reenter self-assembly. However, enzymatic cleavage also leaves behind a stock of residues too large to be incorporated and metabolized by bacteria but too small to self-assemble. These leftovers from bacterial enzyme cleavage are likely to join the refractory organic stock found in the ocean. Implicit in this model is the hypothesis that MG is necessary and perhaps sufficient to drive the global flux of carbon up the food web with ramifications that scale to global cycles of marine bioactive elements.

Results of this work often diverge from established notions prevalent in MG studies, including "exudation" of MBP by phytoplankton [5,6]; studies of MG by techniques borrowed from histochemistry and outcomes measured xanthan gum units [7]; explanations of MBP aggregation based on Smoluchovski's seminal work [8]—which give an excellent account of particle aggregation but tells very little about the nature and mechanisms of supramolecular MBP interactions that result in MG formation—and from explanations of the burning question of why refractory MBP turns inaccessible to bacterial nutrition [9].

Specifics of each of these different lines of inquiry are only briefly presented here, detailed information can be found in the corresponding references. Except, however, in the discussion of unpublished results, previously presented at meetings, which are accompanied by short outlines of the corresponding experimental protocols.

2. Polymer Gel Phase Transition in Phytoplankton Secretion

2.1. First, Some Fundamentals of Polymer Gels

Polymer gels consist of a polymer network and a solvent. While the network entraps the solvent, the solvent keeps the network expanded. Polymer gels hold a highly hierarchical supramolecular architecture in which the polymers in the gel matrix make a 3D network interconnected by chemical or physical cross-links that keep these chains in a statistically stable close neighborhood [10]. Properties of gels—particularly those made of multiple polymer species—can not be explicitly traced to their component polymer chains, or unequivocally predicted from the chemical or physical properties of their polymer components. Complex multiscale systems like MG hold a hierarchy of molecular order in which the role of individual polymers on the properties of the whole network is strictly conditioned by their mutual association with the supramolecular gel matrix.

The combined chemical and physical features of the polymer of matrix—including average chain length, their polyelectrolyte properties, presence of hydrophobic, cationic, or anionic dissociable groups, linear or branched structure, etc.—and the nature of their interconnections establish the topology, chemical reactivity, and bulk physical properties of gel networks and how gels interact with solvents, smaller solutes, and microorganisms [10–16]. Covalently cross-linked chemical gels have a finite limited swelling volume. They do not get interconnected with other gels because polymers are tightly bound to each other and cannot move out of the network to interpenetrate neighboring gels and anneal forming larger gels. On the other hand, physical gels are randomly woven entanglements of chains that are weakly stabilized by low energy bonds, allowing polymers to axially slide past each other. The assembly-dispersion dynamics of tangled networks depend primarily on the second power of the ensemble average of chain lengths of the polymers that make them [11]. Depending on the osmotic balance and density of crosslinks and tangles, these gels can eventually swell indefinitely, and small shear forces can readily disperse their matrix allowing polymers to come apart and dissolve. Conversely, axial mobility of tangled networks can, as well, allows polymers from neighboring gels to interpenetrate and anneal to form larger gels, (Figure 2 and Figure 15) [11,12].

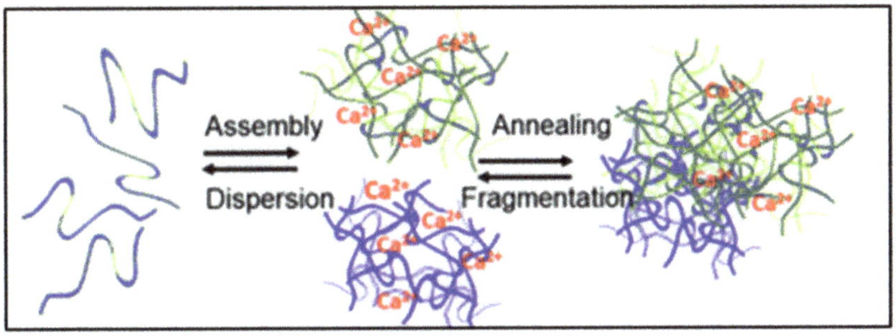

Figure 2. In MG formation are two equilibrium constants: one representing the reversible mass transfer between free polymers and nano-assembled networks, and another that report the dynamics of reversible interconnection resulting from reptation and interpenetration of polymers that result in the formation of MG [2].

Just a few features, out of a broad range of properties of polymer dynamics, are important to understand the way these large molecules are stored inside and released from phytoplankton. One is that they are physical gels and their fate following release from the cell is explained by de Gennes's theory of reptation of physical tangled networks [11]. Second, that their storage and release from the cell is governed by Dušek [17], and Tanaka's [18] theory of polymer gel phase transition. This is a fascinating feature of polymer-gels whereby they can undergo abrupt transitions from a flexible and highly permeable swollen phase to a dense collapsed phase, where the matrix expels most of the solvent—water in hydrogels—and collapse forming a virtually solid particle. This process is reversible and transition from condensed to solvated phase is—as described later—the physical principle that governs the release of MBP in phytoplankton secretion [19,20]. Polymer gel phase transition has the characteristic high cooperativity of critical phenomena and a well-defined critical point. A remarkable feature is that a collapsed gel despite containing a highly concentrated mass of molecules functions as a single supramolecular particle with negligible osmotic activity, which, as far as storage in membrane-bound intracellular vesicles is concerned, makes condensation a very economic storage system with negligible osmotic trans-vesicular osmotic drive [21,22]. Polymer gels phase transition was theoretically predicted by Dušek in 1968 [17] and experimentally confirmed by Tanaka ten years later [18]. However, nature has taken advantage of polymer gel phase transition for a much longer time. Phytoplankton, most likely among the early newcomers in evolution, are equipped with an exocytic mechanism—as sophisticated as the one found in human's secretory cells—and in which polymer gel phase transition plays a critical role [21].

2.2. *Phytoplankton Secretion*

Photosynthesis is the leading source of MBP. About half of the total recapture of atmospheric CO_2 relies on phytoplankton and cyanobacteria, the two main marine photosynthetic agents of the ocean. On average, ~0.5 Gt $\times$ year^{-1} of organic carbon is released to the SW by these microorganisms. The output of this gigantic photosynthetic bioreactor generates an influx of MBP that enters directly or indirectly into the world ocean making about half of the pool of dissolved organic matter; the leading source of nutrients for marine microorganisms [23].

However, despite their critical role in the heterotrophic cycle, the cellular mechanism whereby phytoplankton export organic material to the seawater has remained strictly speculative. Early morphological observations prompted Aaronson to suggest that phytoplankton might indeed function as secretory cells [24]. Instead, a set of unsubstantiated theories proposing that macromolecules are exported across phytoplankton cell membrane by "exudation" or passive diffusion via imaginary ad hoc channels [5,6] are still undisputedly repeated in marine science literature and remain in textbooks as the established paradigm of phytoplankton MBP release to the SW.

The first demonstration that phytoplankton function as secretory cells was conducted in *Phaeocystis globosa* [19,20]. These observations demonstrated that *Phaeocystis* release of MBP exhibit the typical features found in all secretory cells [21,25,26]. Namely, first, that *secretory granules*, in which MBPs are stored, are indeed present in *Phaeocystis* [19] (Figure 3); second, that blue light is the specific stimulus that prompt exocytosis [19] (Figure 4); third, that the light stimulus is relayed inside the cell by a characteristic transient increase of intracellular [Ca^{2+}] [20] (Figure 5); and fourth, that exocytosed material is a gel that, upon exocytosis, exhibit Tanaka's typical dynamics of polymer gel volume phase transition and swelling of polymer gels [10,27,28] (Figure 6).

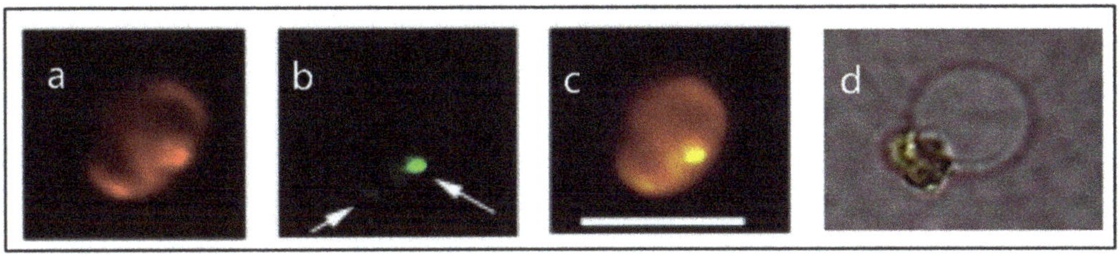

Figure 3. Fluorescence images of *Phaeocystis globosa* taken with excitation at λex = 494 nm wavelength that induces emission of both chlorophyll in the chloroplast, that fluoresces bright red (Panel **a**), and of the secretory granules labeled LysoSensor—a probe that binds specifically to secretory granules—that fluoresces in green. The green granules indicated by arrows (Panel **b**) are almost at the limit of resolution of the optical microscope. A few can be distinguished individually, and most of them—like in this case—are clustered inside the cell. Panel (**c**) is the merged image of (**a**) and (**b**). Depending on [Ca^{2+}] and pH in SW, the dimensions of the swollen exocytosed MBP matrix (Panel **d**) can vary from ~3–12 μm. Bar = 8 μm.

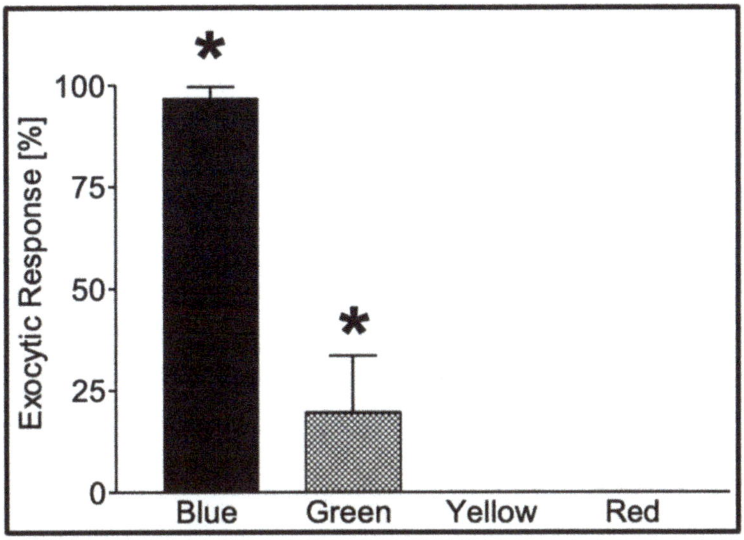

Figure 4. Secretion in *Phaeocystis* responds to specific photo-stimulation. Exposure of *Phaeocystis* to a photon flux (*) of 480 μmol $m^{-2} \times s^{-1}$ of blue light (λ = 450–490 nm) for 60 min induced secretion in 97 ± 2.9% of the cells, while similar exposure (*) to green light (λ = 500–540 nm) resulted in secretion in only 20 ± 14% of the cells. Exposure to equal time/fluxes of yellow (λ = 560–600 nm) and red (λ = 640–720 nm) light fails to induce secretion. Bars correspond to the mean ± SEM of five measurements [17].

Secretion in *Phaeocystis* responds to specific photo-stimulation triggered by blue light, and to a lesser extent by green light [19] (Figure 4).

A characteristic transient increase of intracellular $[Ca^{2+}]_C$ takes place following stimulation in all secretory cells [21,29]. We studied Ca^{2+}-signaling in photostimulated *Phaeocystis* by using the membrane-permeant fluorescent Ca^{2+} probe Fluo 4-AM. As shown in Figure 5, exposure of *Phaeocystis* to 450–490 nm blue light resulted in a characteristic increase of intracellular $[Ca^{2+}]_C$ that was consistently followed by exocytosis [20].

Video recordings show that during *Phaeocystis* exocytosis the granule's polymer matrix undergoes the typical swelling that characterizes the transition from condensed to hydrated phase of polymer hydrogels [19,27,28]. Depending upon [Ca^{2+}] and pH of SW, the radial expansion of the exocytosed gel can increase from ~1 μm to up to ~3–10 μm. It follows the typical features of swelling of polyelectrolyte gels [19,27,28]. This process can be formalized and evaluated according to Tanaka's theory of swelling of polymer gels [27] to infare the diffusion coefficient of the MG's matrix network [28].

Figure 6 illustrates the swelling kinetics in a *Phaeocystis'* exocytosed granule's cargo. Measurements conducted by digitizing video microscopy recordings show that the radius of *Phaeocystis'* secreted polymer gels increases following typical first-order kinetics. The continuous line is a non-linear least-square fit of the data points to $r(t) = r_f - (r_f - r_i)e^{-t/\tau}$, where r_i and r_f are the initial and final radius of the granule, and τ is the characteristic relaxation time of swelling. As in other secretory cells [28], the final radius of r_f of the exocytosed gel shows a typical linear relationship with τ^2, the second power of the characteristic time of swelling (Figure 6).

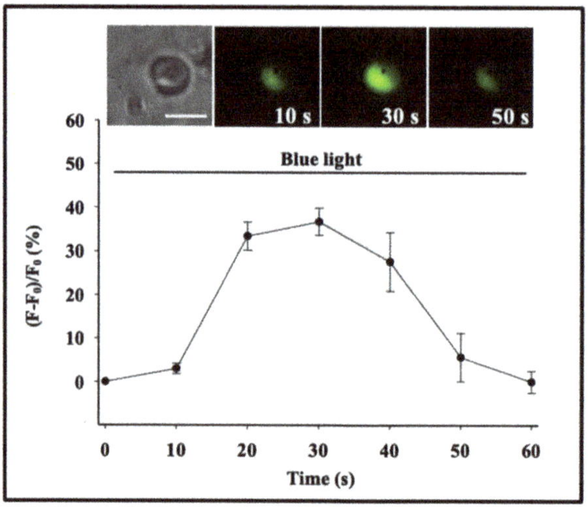

Figure 5. A characteristic transient increase of intracellular $[Ca^{2+}]_C$ takes place following stimulation in *Phaeocystis* [20]. We studied Ca^{2+}-signaling in photostimulated *Phaeocystis* by using the membrane-permeant fluorescent Ca^{2+} probe Fluo 4-AM. As shown here, exposure of *Phaeocystis* to blue light stimulation (λ = 450–490 nm) results in a transient increase of intracellular $[Ca^{2+}]_C$ reported by a corresponding increase of fluorescence expressed as the ratio of emission before and after stimulation. This typical increase of $[Ca^{2+}]_C$ was consistently followed by exocytosis and release of MBP stored in the secretory granule. Each point corresponds to the mean ± SEM of nine measurements. Bar = 10 µm.

The slope of this line $\tau = D\,(r_f)^2$ represents the diffusivity (D) of the granule secretory matrix in SW (Figure 7). As predicted by theory, and observed in other bio gels, the diffusivity of swelling polyelectrolyte hydrogels, like those that make the matrix of secretory granules, follows Donnan equilibrium [30–33] and depends on the counterion concentration in the swelling medium. Figure 7 illustrates the effect of $[Ca^{2+}]$ concentration in SW on the diffusion of *Phaeocystis* exocytosed gels [19]. D is also affected by SW pH as well as the presence of higher valence cations often found as pollutants in marine habitats.

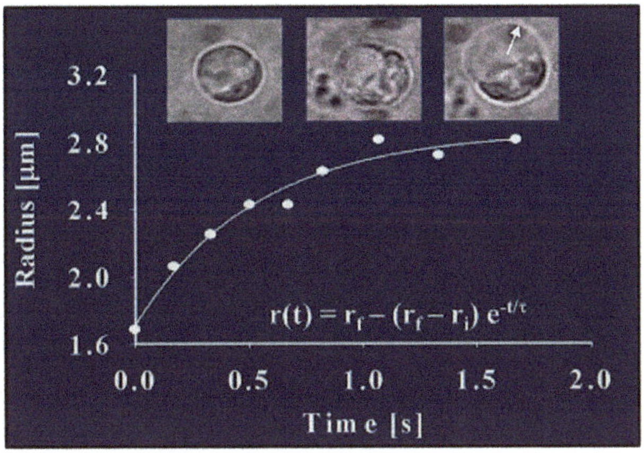

Figure 6. The radial expansion (arrow) of the exocytosed granules follows characteristic first-order kinetics, lending the process to be formalized and evaluated in light of Tanaka's theory of swelling of polymer gels [27]. This figure illustrates images of video recording of exocytosis in a *Phaecystis* cell captured at 30 fields $\times$ s^{-1}. Measurements conducted by digitizing video microscopic images show that, during product release, the radius of the secreted polymer gel increases following first-order kinetics. The continuous line is a non-linear least-square fitting of the data points to $r(t) = r_f - (r_f - r_i)e^{-t/\tau}$, where r_i and r_f are the initial and final radius of the granule, and τ is the characteristic relaxation time of swelling.

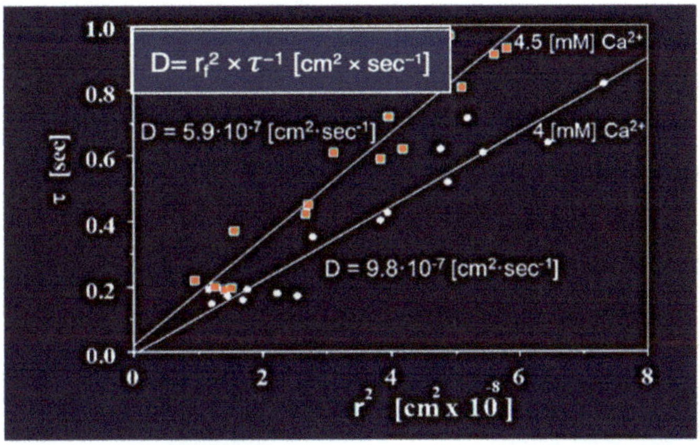

Figure 7. In here, *Phaeocystis* was equilibrated in ASW containing 4 mM and 4.5 mM $[Ca^{2+}]$ at pH 8.2, 20 °C, and stimulated to secrete by exposure to blue (450–490 nm) light. The swelling of exocytosed granules was monitored by video microscopy. In agreement with theory [25], the relationship between characteristic time (τ) of the swelling kinetics, and the second power of final radius $(r_f)^2$ of the exocytosed granules exhibit a characteristic linear function with a dimension of $cm^2 \times s^{-1}$ that represent the diffusivity $D = (r_f)^2 \times \tau^{-1}$ $[cm^2 \times s^{-1}]$ of the exocytosed gel. Counterion concentration affects the diffusivity of exocytosed gels. In here *Phaeocystis* slight changes of concentrations of ASW $[Ca^{2+}]$ from 4 mM to 4.5 mM at pH 8.2, 20 °C, result in a corresponding decrease of D, from 9.8×10^{-7} $[cm^2 \times s^{-1}]$ to 5.91×10^{-7} $[cm^2 \times s^{-1}]$.

As expected, the effect of increasing $[Ca^{2+}]$ in ASW illustrated in Figure 8, results in a power-law decrease of the diffusion of the exocytose *Phaeocystis* gel.

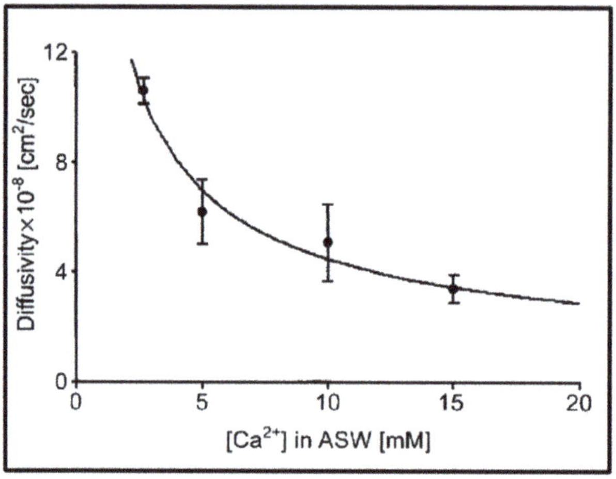

Figure 8. Experiments conducted in *Phaeocystis* equilibrated in artificial seawater (ASW) containing increasing concentrations of Ca result in a characteristic power-law decrease of the diffusion of the exocytose *Phaeocystis* gel. Point corresponds to the mean ± SEM of seven measurements.

As shown in Figure 9, the hydrated polymer matrix of exocytosed *Phaeocystis* gels when exposed to conditions that mimic the intragranular environment, i.e., low pH and high [Ca^{2+}] can recondense undergoing a characteristic polymer gel volume transition [22].

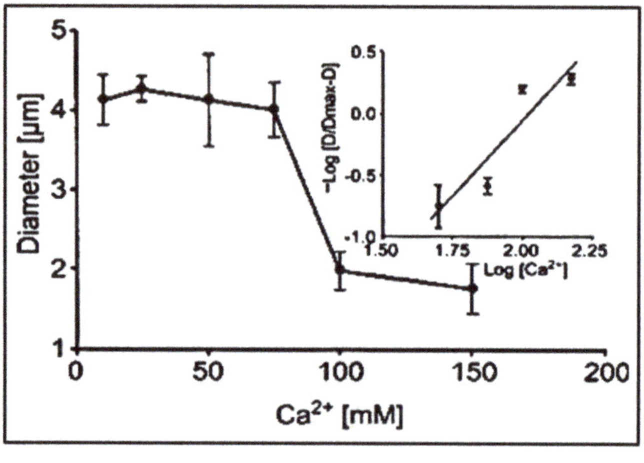

Figure 9. The diameter of swollen exocytosed gels—equilibrated in a pH 3 buffer solution—changes drastically as [Ca^{2+}] increased from 10–150 mM $CaCl_2$. It undergoes a characteristic transition to a condensed phase [19]. It shows a steep inflection at a critical point between 60 and 100 mM $CaCl_2$ at which phase transition takes place. The process is reversible, and as shown by the Hill plot in the inset, it exhibits the characteristic high cooperativity of critical phenomena with a Hill coefficient >2.5. Point corresponds to the mean ± SEM of seven measurements.

In summary, these results show that *Phaeocystis* is indeed a secretory cell. It holds granules that contain a typical polymer gel that remains in condense phase during storage undergoing volume phase transition to solvated phase upon release from the cell. Counterion dependence and high diffusion of *Phaeocystis* exocytosed networks indicate that it is driven by the fix charges of MBP polyanions and governed by a Donnan equilibrium [30–33]. The fact that, at long swelling times, *Phaeocystis* exocytosed networks can disperse suggests that the matrix of phytoplankton swollen hydrogels has a loosely woven random tangled topology. In these physical networks, the translational diffusion of the polymer chains in the matrix is constrained to a snake-like axial motion through the inter-twining of surrounding polymer molecules [11]. This is known as reptational axial diffusion [12]. It allows polymer chains to move out of the matrix and disperse, or to interpenetrate adjacent gels interconnecting them together (Figure 2). Thus, a tangled topology of the polymer matrix of *Phaeocystis* granules explains how these exocytosed microscopic gels can anneal together forming the characteristic large masses of mucilage found during blooms, or else disperse to join the DOM pool of the ocean (DOM is operationally defined as all organic moieties found in 0.2–0.7 µm pore size the filtrate of SW).

Immunocytochemical detection in SW collected at stations located in a wide geographical distribution shows that MBP released by *Phaeocystis* is broadly found throughout the water column [34]. It suggests that *Phaeocystis* could not only be an important contributor to the global DOM [23] and MG stock, as the MBP it releases can eventually self-assemble forming microscopic MG [2].

2.3. Phytoplankton Toxin Release

Another significant feature of polymer dynamics in secretion is that the granule's matrix of secretory cells regularly cages active products that the cell releases. For instance, in chromaffin cells, the matrix is made from chromogranin—a strong polyanionic polymer—and the active product exported is epinephrine; in mast cells, the matrix is made of heparin, and the active product these cells export is histamine, in goblet cells the matrix is made of mucins and the active product are antimicrobial peptides called porins [21]. Unpublished observations illustrated in Figures 10 and 11, show that a similar mechanism of storage and release is probably present in the dinoflagellate *Karinia brevis*. *K brevis* is a toxic dinoflagellate responsible for red tide outbreaks throughout the world. Dangerous outcomes of these blooms are caused by brevetoxin, a potent neurotoxin responsible for substantial marine life mortality and human morbidity.

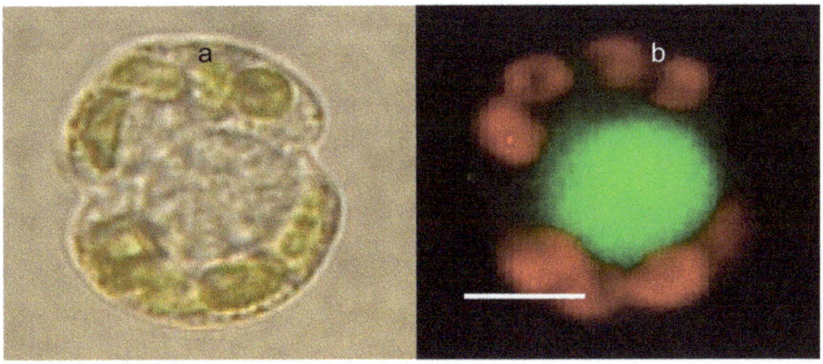

Figure 10. Phase contrast (**a**) and LysoSensor-labeled *Karinia brevis* (**b**). The red emission corresponds to the autofluorescence of Chlorophyll from Karenia's chloroplasts. The green emission is from a large secretory granule, labeled with LysoSensor, a probe that binds specifically to secretory granules. Bar = 8 µm.

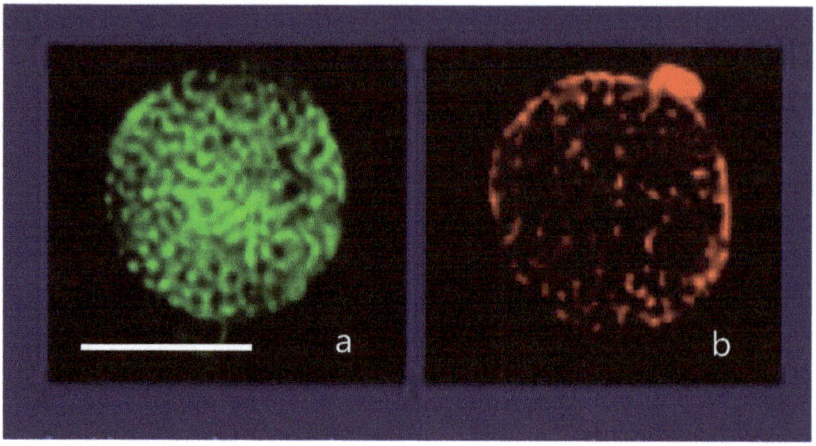

Figure 11. Confocal image of a secretory granule of *Karenia brevis* demonstrating colocalization of LysoSesor green—that labels secretory granules—in (**a**) and a fluorescent anti-brevetoxin antibody conjugated with Tetramethyl Rhodamine Isothiocyanate that emits in red (**b**). Bar = 10 µm.

The mechanisms whereby *Karinia* releases toxins to the SW remained unknown, explained by the imaginary untested doctrine of exudation. Fluorescence microscopy images of *K. brevis* double-labeled with LysoSesor green—that label secretory granules—and fluorescent anti-brevetoxin antibody show that brevetoxin colocalizes inside *K. brevis* granules. It is, like in other secretory cells, caged inside the condensed matrix *K. brevis* secretory granules (Figures 10 and 11).

Exposure of *K. brevis* to blue light for 60 s can readily stimulate exocytosis in these cells. These observations indicate that *K. brevis*, like *Phaeocystis*, is a secretory cell; that brevetoxin is stored in secretory granules and is released by exocytosis following photo-stimulation by blue light.

3. Marine Biopolymer Self-Assembly

Advances in marine geochemistry render a remarkable complexity of organic polymers dissolved in SW. These biopolymers hold a huge mass of reduced organic carbon reaching more than 700 Gt. At the micromolar concentrations found in SW, MBPs are unlikely to undergo significant chemical interactions. Conversely, because of their polyanionic nature, these chains can fully interact with SW metal counterions to form cationic bonded supramolecular networks that make the matrix of physical gels.

However, the broad range of chemical species present in DOM makes it extremely intricate to arbitrarily specify structure-function assignments to predict the complex features that result in the formation of gels in seawater (SW). DOM chemistry makes up a body of excellent science but only a fraction of it can serve to reliably—and even then, to only partially—forecast the dynamics of MG formation. However, relevant physical information—about low energy macromolecular dynamics—required to predict specific roles in gel formation, like their contour length, Z potential, polyelectrolyte properties, hydrophobic-hydrophilic features, are largely still missing.

On the other hand, there is well-tested theory that sets the laws that govern polymer gel dynamics [11–15,17,18]. However, experimental validation of these theories has been mostly conducted in simple synthetic chemically crosslinked gels, or physical gels made of monodisperse polymers solutions, both conditions far simpler than the complex polydisperse scenario of assembly and gel formation taking place in SW. In short, we are still at the start of harvesting the full predictive power of physical theory to crack the complex and exciting riddles hidden in the sea.

Marine gel formation has been successfully described and modeled using Smoluchovski's aggregation theory [8]. However, while this approach can effectively and elegantly account for DOM aggregation and MG formation, it gives no inside into the molecular mechanisms whereby MBPs associate to form gels. Behind the concept of stickiness hide a black box, with plenty of room for speculation, that tells little about what specific molecular interactions are at play in marine gel formation.

In short, the ocean is the second most important Carbon cycling reactor on our planet and MGs are likely to be a critical conduit between source and sink in this process (Figure 1). If this hypothesis turns out to be correct, it makes MG a central figure in the critical role of the ocean in global carbon cycling. The discovery that carbon in DOC—the component of DOM quantifiable as total organic carbon—can be reversibly transferred to carbon in MG, (MGOC)—the component of MG quantifiable as total organic carbon—via MBP self-assembly [2] implies that a corresponding reversible DOC↔MGOC equilibrium must exist. Thus, an objective estimate of carbon flux going through the MG pool can give a valuable parameter to gauge the rate of the source to sink carbon flux in the ocean (Figure 16). To do so requires reliable measurement of MGOC. Unfortunately, however, MGs remain investigated using qualitative methods borrowed from medical histology. Accordingly, gels are detected, and their concentrations are estimated by using multiple different versions of a qualitative indirect histochemical technique whereby gels are measured in gum equivalents, and carbon content is expressed about xanthan gum carbon [35–40]. As expected, there are broad inconsistencies of results among these different colorimetric assays. Discard et al., [41] conducted the only systematic evaluation comparing the multiple colorimetric assays that have been introduced following the pioneering work of Aldredge and Passow [8,35]. They found that all these methods fail to deliver absolute reliable figures of gel concentration in SW, and, by implication, fail to render objective quantitative measurements of MGOC. They conclude that it is virtually impossible to compare results among laboratories using these different techniques. Nonetheless, it is important to emphasize the decisive role that Aldredge and Passow's work has played in giving recognition to the crucial significance of the multiple complex functions that MGs play in the world ocean. Despite their limitations, these assays continue to render a broad phenomenology with thorough descriptions, classifications, correlations, and modeling of gel dynamics in the ocean. Unfortunately, however, these colorimetric methods fail to render absolute figures of MG gel concentration and robust understanding of the fundamental physical mechanisms whereby MGs are formed or dispersed, their Donnan ion exchange and chemical partition properties, their phase transition features, etc. Moreover, as far as the central role of MG in ocean carbon cycling is concerned, the principal limitation of these colorimetric assays is that they fail to objectively measure MGOC, a parameter that together with DOC allows estimating the input-output dynamic-flux of carbon passing from source to sink via the MG phase of the ocean (Figure 16). A simple dye-free direct assay to directly measure MGOC was introduced by Chin et al. [2], and it is outlined in the Appendix A.

3.1. Polyelectrolyte Marine Gel Networks

The ocean probably holds an unknown stock of covalently crosslinked biopolymers found in the remains of dead cells and tissues. Those are chemical gels whose turnover and role in carbon flux remain as a task for the future. However, gels resulting from interactions of biopolymers found in the DOM pool are physical gels. In these gels the matrix is interconnected by tangles and low-energy physical bonds, forming a three-dimensional random network. Gel size in this case results from a reversible turnover of assembly/dispersion equilibrium in which cross-links and tangles are continuously being made and broken. Polymers in solution move by diffusion or convectional drive and depending on their concentration, flexibility, and length they can form inter or intrachain transient tangles. Because of their polyelectrolyte properties, if pairs of ionized groups approach the Coulomb field of a polyvalent counterion, a reversible crosslink can be formed. The binding energy of electrostatic bonds is low, they continuously and randomly switch from bound to unbound. The stability of polyelectrolyte physical networks is at the crossroads of two exponential functions, i.e., it depends on the second power of polymer chain length [11,12], and the second power of the valence of the crosslinking counterion [42]. Thus, the average length of MBP in the MG's matrix, the number of charged groups in these chains, the number of bonds attached at any time, and most importantly, the valence and concentration of crosslinking counterions are all critical for MBP self-assembly and MG formation.

A similar outcome takes place when hydrophobic domains in diffusing amphiphilic polymers approach each other. Hydrophobic connections are highly dependent upon temperature and require short interchain distance thereby making them highly dependent on polymer concentration; a condition that is particularly relevant at the marine air-water interphase where amphiphilic moieties concentrate with their hydrophobic heads buoyant and hydrophilic tails immersed. The ratio of hydrophobic/hydrophilic domains of MBP, their concentration, and the concentration of short-chain-crosslinker ampholytes in SW are important determinants of hydrophobic bond formation. Electrostatic, hydrophobic, or hydrogen bond, interconnections need little activation energy (<50 kJ mol^{-1}) and are fully reversible, with bonds continuously and randomly making and breaking. Thus, the stability of physical gels is like the story of Gulliver tied to the ground: it relies on the presence of many weak attachments, and on the fraction of bonds that remain locked at any time.

3.2. *MBP Is the Feedstock of Marine Gels Formation*

MBP forms an assorted set of macromolecules that include mostly aliphatic unbranched polysaccharides, proteins, nucleic acids, and lipids. They are largely polyanionic often amphiphilic chains of different sizes [42–45]. However, the bonding that interconnects them in MGs belongs to only four categories of low energy interactions. Namely, electrostatic links, hydrophobic bonds, hydrogen bonds, and tangles. Of those, only electrostatic and hydrophobic bonds have been experimentally evaluated [2,46,47].

Detailed chemical composition of MG has not been established but labeling with specific fluorescent probes indicates that they are made not only from primary production exopolymers but of a complex mix of biopolymers including polysaccharides, proteins, lipids, and nucleic acid residues of multiple origins (Figure 12).

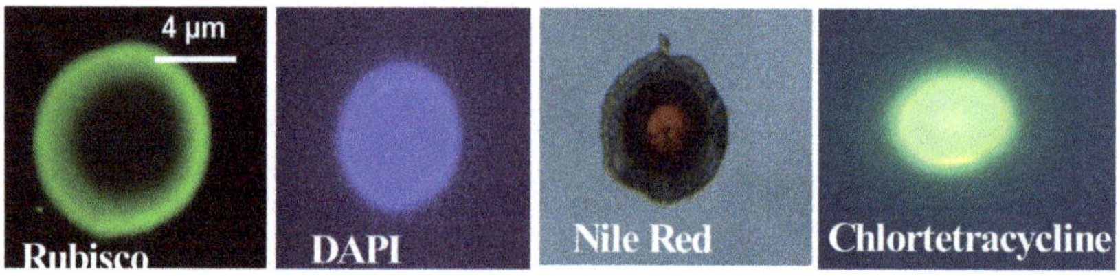

Figure 12. Fluorescence images show that marine gels contain a broad variety of biopolymers, including rubisco—a protein, labeled with a fluorescent specific antibody—nucleic acids—labeled with DAPI—hydrophobic moieties, probably lipids—labeled with Nile Red—and polyanionic polymers that bind Ca—labeled with CTC (Chlortetracycline).

3.2.1. Dynamics of Ca Crosslink-MBP Self-Assembly

The first objective evidence that DOM biopolymers self-assemble forming microscopic gels was reported in 1998 [2]. MBP self-assembly follows characteristic second-order kinetics (Figure 13). It has a thermodynamic yield of ~10% calculated from the difference between carbon—measured by high-temperature catalytic oxidation—found in 0.2 μm-filtered SW samples conducted immediately after filtration—before gels are formed—and 0.2 μm-filtered SW samples incubated for 150 h allowing MBP to self-assemble and refiltered at 0.2 μm-filtered—in which case gels are retained in the filter but let pass the DOM that failed to assemble.

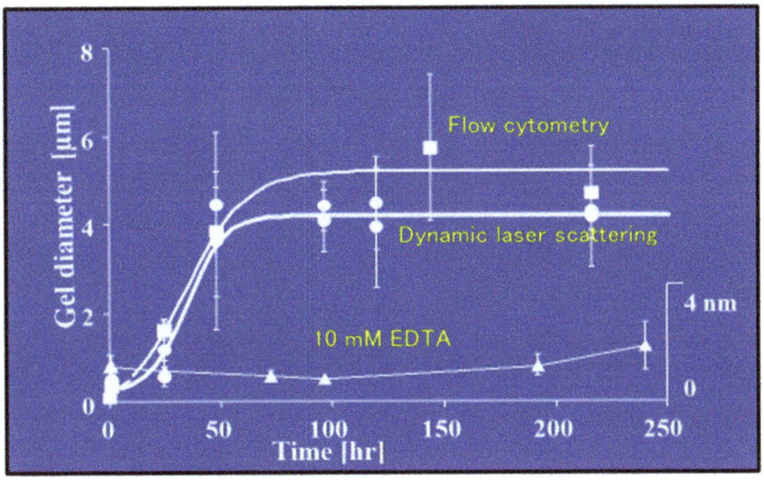

Figure 13. The Time course of DOM biopolymer assembly and MG formation is monitored by measuring particle size by both homodyne dynamic laser scattering (DLS), and by flow cytometry in 0.2 µm-filtered SW. Measurements of microgel size by flow cytometry (squares) and DLS (circles) show a similar time course of assembly. MG size grows from colloidal, submicrometre size, to several micrometers following characteristic second order kinetics. Control samples in which MBP assembly was inhibited by chelating Ca^{2+} by adding 10 mM EDTA to the SW gave an average size of 1–2.5 nm, regardless of the time of observation (triangles). Each point corresponds to the average ± SD. of five measurements.

This simple assay yields robust direct figures—not referenced to gum units—of marine gels' organic carbon (MGOC) content. The thermodynamic yield of gel assembly indicates that ~70 Gt (70 × 10^{15} g) of reduced organic carbon is likely to be present in the ocean as MGOC [2]. This figure is ~1.5 orders of magnitude higher than the total marine biomass estimated at ~4 Gt [48].

Results illustrated in Figure 13 indicate that self-assembly of MBP results from counterion crosslinking where Ca^{2+} divalent cations are inter-connecting MBP chains. Experiments where—instead of chelating counterions—SW was dialyzed against Ca-free artificial SW, reported similar self-assembly kinetics (not shown), indicating that MBP assembly stems principally from Ca^{2+}, not Mg^{2+} ion crosslinking. The absence of crosslinking by Mg^{2+} might stem from polymer-cation affinity due to the different sizes and shapes of the hydration shells between these two cations. This outcome is confirmed by measurements of the elemental composition of MG using electron probe microanalysis that indicates a high level of Ca but low levels of Mg content (Figure 14). Assumptions about Mg operating as a counterion crosslinker in MGs assembly have not been experimentally demonstrated. Published evidence [2] as shown below clearly indicate rule out this speculative outcome.

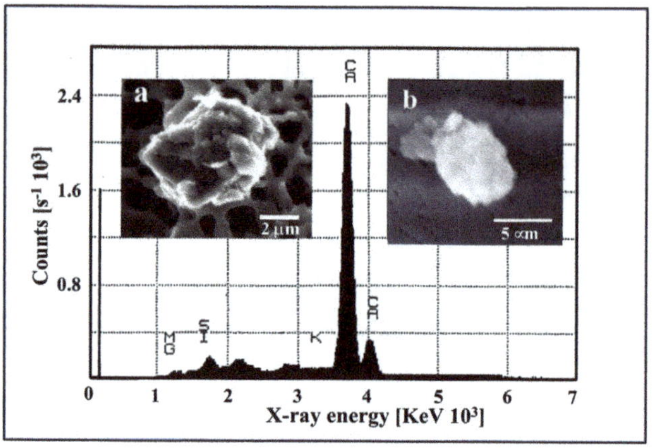

Figure 14. Electron probe microanalysis reveals that MGs contain substantially higher levels of Ca than Mg. Notice, that probably resulting from the Donnan partition, Ca reaches much higher concentrations inside self-assembled MG than in SW. Inset (**a**) is an environmental scanning electron micrograph image of a MG, the background depicts the surface of a filter—notice that this gel is intact and fully hydrated as environmental scanning electron microscopy (ESEM) does not require fixation or any other artifact-inducing chemical manipulation. Inset (**b**) is a gel labeled with CTC, a fluorescent probe that labels bound Ca and that, unlike other colorimetric probes, does not produce crosslinking of MBP or affect in any way the supramolecular structure of gels.

The second-order kinetic illustrated in Figure 13 indicates that there is more than one step in MBP self-assembly. We conducted objective verification of intermediate steps of assembly by imaging material found in aliquots at the start, and after 4 h, and 24 h of assembly. Atomic force microscopy (AFM) imaging and environmental scanning electron microscopy (ESEM)—both methods that do not require "fixation" or staining of samples—were used in these experiments shown in Figure 15. Notice that the background in the ESEM picture (right panel) corresponds to the surface of the filter.

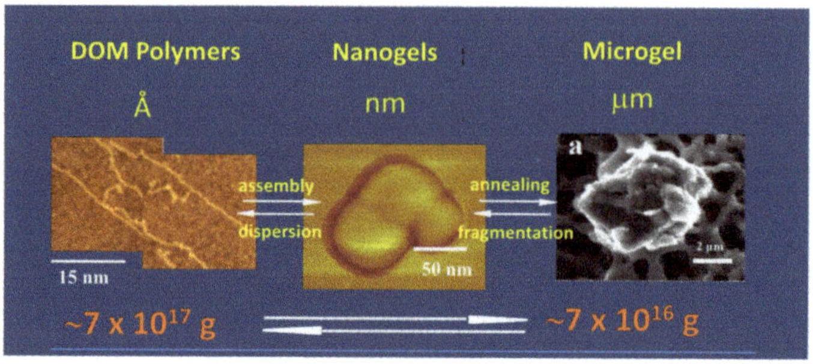

Figure 15. This figure shows that the formation of MG starts from biopolymer precursors at Angstroms dimensions (**left panel**). They self-assemble forming nanogels (**center panel**) that anneal to final equilibrium gel size at micron scales (**right panel**). MBP and nanogels were set on mica substrate for ATM imaging. A filter surface can be seen in the background of the ESEM image on the right. Arrows indicate material flow between MBP, nanogels, and micron-size MG, and in the lower panel illustrate organic carbon mass equilibrium between DOM and MGs.

A simple two-step kinetics model of MBP assembly is illustrated in Figure 16.

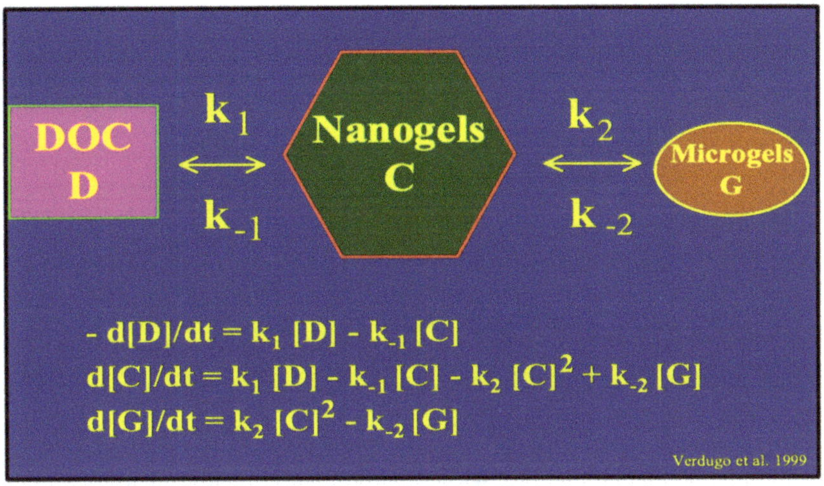

Figure 16. A simple second-order kinetics mathematical model of MBP assembly can be formulated by assigning two correspondent equilibrium constants k_1 and k_2 to the assembly reaction. Square parenthesis refers to Carbon concentration in each of the subsequent steps from DOC to Nano and MGs. From the MG step, organic carbon becomes available to microorganisms, and flux becomes irreversible as carbon enters the food web escalator (Figure 1). Notice that MGOC can be a sensitive indicator of the balance between supply and demand of carbon in marine carbon cycling.

The demonstration that electrostatic interactions play a critical role in MBP crosslinking indicates that other polycations might as well result in MG formation in SW. As pointed earlier counterion crosslinking of polyelectrolytes is complex and dependent upon several factors. Paramount among those is that other variables being equal, crosslinking is proportional to the second power of the valence of the crosslinking counter ion [45]. Polluting heavy metals can affect MG assembly at very low concentrations. For instance, Al^{3+}—which is used routinely in sewer water treatment—can induce MBP association and aggregate formation at micromolar concentrations. Another polycation that requires attention is Alcian Blue. This is a strong polyvalent basic dye with four cationic sites that can readily bind and crosslink polyanions, including of course MBP. A question that needs to be experimentally evaluated is if Alcian Blue, which has been applied routinely to study MGs—called transparent exopolymer particles (TEP)—might itself produce MGs. This is an issue that requires close and rigorous examination as it might explain why different Alcian Blue assays—using different filters, different Alcian Blue concentrations, or different pH—that can induce MG phase transition—yield different results [35–40] affecting the significance of a large body marine gels work [41].

However, the issue that requires most close monitoring at this time is the changes in SW temperature and pH, both variables critically affected by global warming which is already driving a progressive deterioration of marine life. SW pH and temperature can certainly affect the kinetics and thermodynamic of the MBP association. As the world ocean acidifies, MBP carboxylic groups—pKa~4.5—which are the residues that allow MBP to associate, get progressively protonated potentially decreasing polyanion-Ca affinity and crosslinking dynamics. On the other hand, as shown below, increased SW temperature can increase the probability of hydrophobic bonding and interchain association leading to increase MG formation. These are theoretical predictions and homework for young chemical oceanographers.

3.2.2. Hydrophobics Bonds in MBP Self-Assembly and MG Formation

The polymer matrix of MGs contains not only polyanionic polysaccharides—that participate in Ca-crosslinking—it also contains proteins potentially having both ionizable and hydrophobic groups, as well as lipids, which are strong hydrophobic polyampholytes (Figure 12). Thus, there is plenty of opportunities that hydrophobic interactions might be at work in MBP self-assembly. We explored the role of hydrophobic interactions in MG formation in collaboration with Peter Santschi's group. Among the potential candidates to crosslink MBP are moieties called exopolymer substances (EPS) produced by bacteria. These amphiphilic molecules are thought to induce the formation of particles in SW, Decho [49] and Stoderegger and Herndl [50] first introduced the notion that the gel-inducing properties of EPS are related to their relative hydrophobicity. However objective validation of this hypothesis was only recently conducted [46,47].

EPS from *Sagittula stellata* was purified at Santschi's laboratory. We determined its hydrophobicity by fluorescence energy transfer using Sagittula EPS (SEP) as a donor fluorophore and Nile Red as acceptor chromophore. Fluoresce energy transfer indicated that, among its multiple components, SEP must contain, at least one moiety that is strongly hydrophobic [47].

Results of measurements of self-assembly using DLS in Figures 17 and 18 show that micromolar concentrations of SEP in SW can both induce self-assembly of SEP and can induce the assembly of MBP. These features closely resemble the assembly of polymer-surfactant cosolutes: the assembly follows a first-order kinetics and requires a much lower concentration than the critical assembly concentrations of a polymer or surfactant alone [51]. Although de Gennes's theory of polymer solvation in mixed good solvents close to a critical point [52] offers a simple qualitative model to explain polymer-surfactant assembly the fundamental mechanisms of hydrophobic bonding remain unclear.

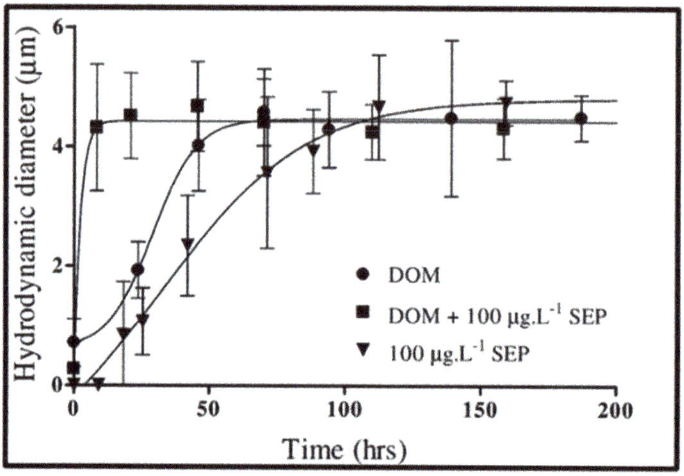

Figure 17. Assembly kinetics of DOM polymers in 0.2 µm-filtered seawater (circles) and self-assembly of SEP (100 µg × L^{-1} in ASW (triangles) monitored by DSL. Addition of 100 µg × L^{-1} to 0.2 µm-filtered seawater results in quick DOM assembly that reaches equilibrium in ~10 h yielding microgels of ~4–5 µm hydrodynamic diameter (squares). Points are the average ± SD of 15 outcomes of triplicate measurements in five samples.

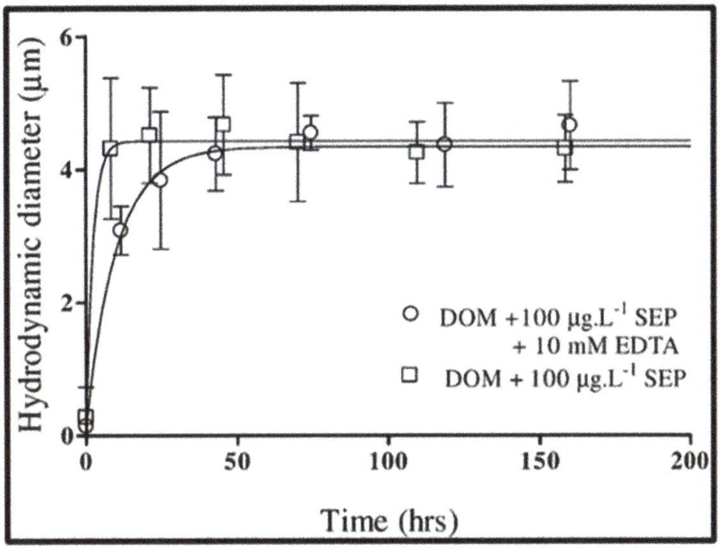

Figure 18. Kinetics of self-assembly of DOM polymers induced by adding 100 µg × L^{-1} SEP to 0.2 µm-filtered seawater in the absence (open circles) or presence of 10 mM Ca^{2+} chelator EDTA (open squares) does not exhibit significant statistical differences. Points are the average ± SD of 15 outcomes of triplicate measurements by DSL in five samples.

It is important to emphasize that in both instances illustrated in Figures 17 and 18, assembly takes place in absence of Ca^{2+} ions. Notice as well that in this instance, the time course of assembly departs from the second-order kinetics observed in Ca-crosslinked MBP assembly: it follows fast first-order kinetics suggesting that hydrophobic crosslinks—much stronger than Ca bonds—must leave no room for reptational diffusion, interpenetration, and annealing of nanogels. Instead, MBP gets progressively locked on the network until their hydrophobic bonds are balanced by shear forces that shave polymers away from the assembled net. Thus, while mechanisms remain obscure, the phenomenology of hydrophobic MG self-assembly is clear.

Although these assays were designed to demonstrate MBP self-assembly via hydrophobic bonding, the results open another intriguing implication. Namely, it portrays a scenario whereby bacteria may release a chemical snare that at micromolar concentration can readily lock, immobilize, and concentrate organic substrate at a close neighborhood, i.e., MB may be able to surround themselves inside a nutritious self-generated gel. These observations suggest that for bacterial nutrition, EPS may be a reagent as important as exoenzymes, and that—among multiple DOM species—amphiphilic MBP might be bacteria preferred bioreactive substrate [16].

Further work conducted in exopolymers released from *Synechococcus*, *Emiliana huxleyi*, and *Skeletonema*, indicates that in all these instances MBP exocytosed by these phytoplankton unicellulars self assembles via hydrophobic bonds [46].

Amphipathic moieties concentrate at the air-water interphase with their hydrophobic heads buoyant and their hydrophilic tail immersed. Thus, the gigantic area of the air-water interphase of the world ocean is likely to be a rich source of marine hydrophobic self-assembly. This unique interphase-driven partition of amphipathic polymers may explain both, the mechanism of formation of gel produced by bubble formation in the laboratory as well as in natural conditions whereby microgels accumulate on the ocean surface and—as recently demonstrated—eventually pass to the atmosphere [53]. Climate change can perturb two parameters that, among other consequences, can strongly influence hydrophobic interactions; namely, SW increased temperature and acidification are both undergoing progressive fluctuations that may strongly interfere with MG formation.

In summary, it is most likely that, given the complex composition of DOM, an equally complex polymer self-assembly kinetics must result from multiple interactions including hydrogen bonding, and both electrostatic and hydrophobic interactions resulting from corresponding charged groups and amphipathic polymers found in SW.

3.2.3. Marine Gels Phase Transition

Like any other polymer gel that knows its table manners, MG can indeed undergo typically reversible volume transitions. Phase transitions are formalized by a typical power-law function like those that describe changes of solid-liquid, or liquid-gas transitions. They all exhibit characteristic high cooperativity and take place at a defined critical point whereby a small change around the intensive value of an environmental variable can trigger a phase change [22]; like, for instance, water changing from liquid to gas phase at precisely 100 °C at sea level pressure. Likewise, changes in pH or temperature can readily condense or decondensed MGs [2]. Figures 19 and 20 illustrate the characteristic high cooperativity of volume change produced by varying temperature and pH of SW.

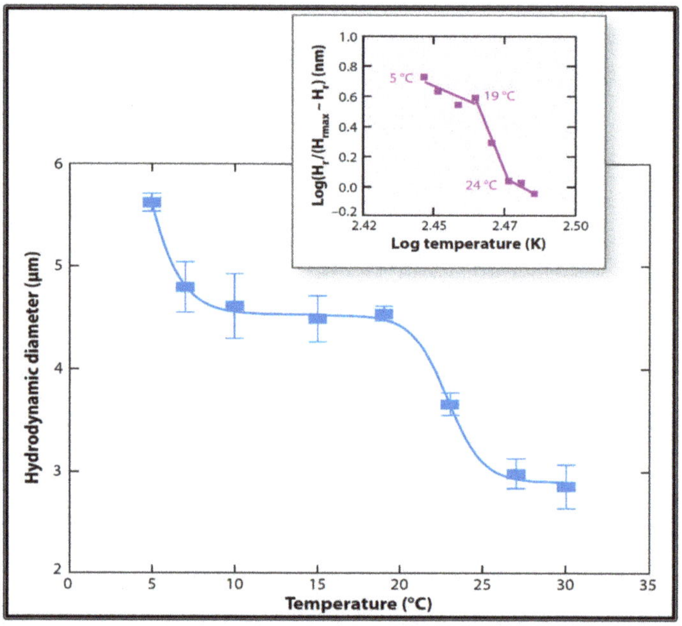

Figure 19. Measurements of MG hydrodynamic diameter by Photon Correlation Spectroscopy show that MG exhibits a two-step temperature-induced phase transition, with critical points at ~7 °C and between 20 °C and 25 °C. Inset. The Hill coefficients in both cases exhibit values that portray the typical high cooperativity of critical phenomena. Each point corresponds to the average ± SD. of five measurements.

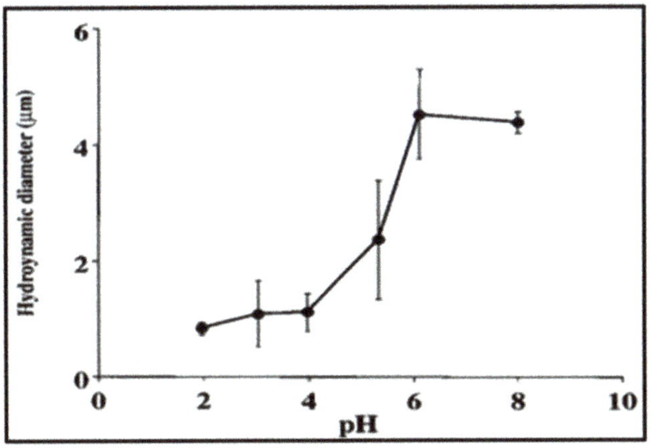

Figure 20. pH-induced phase transition of MG. Notice that this transition occurs within the same range as the expected pKa (4.5–5) of carboxylic groups of DOM. Each point corresponds to the average ± SD of five measurements.

Given the complex macromolecular composition of marine gels, it is not surprising to find broad or multiple critical points (Figures 19 and 20). The significance of these observations is that these transitions occur at ranges of temperatures normally found in the ocean. If we consider that condensation increases the relative density of gels, transitions may eventually increase the sedimentation coefficient of MGs, and the export of condensed networks down the water column. This potential interesting outcome remains as homework for the future.

Conversely, seawater pH at which MG undergo phase transition is probably only found in hydrothermal vents, or perhaps at early times of geological and biological evolution of our planet. Nonetheless, the results in Figure 19 indicate that a low pH, at values like those often used in standard protocols for gel detection [35–40] can readily induce phase transition in marine gels.

Phase transition opens several interesting lines of inquiry. For instance, chemical interactions that may not take place in swollen hydrated networks may occur in a condensed phase. In these tightly packed networks, polymers or other entrapped species become in close contact potentially allowing chemical interactions that otherwise do not occur in seawater, eventually producing species whose origin has not been unequivocally established.

High-pressure phase transition has been demonstrated to occur in synthetic polymer gels [54]. A pending riddle is if high pressure—like it is found in the dark ocean—might induce phase transition of MG. If pressure induces marine gels condensation, the sediment of the ocean may be a collecting site of material largely not accessible to bacteria. It is important to remember, that in condensed networks the Debye-Hückel screened Coulomb potentials and the distance among polymer chains are thought to be collapsed [18]. i.e., there is practically no navigating water inside the collapsed gel for enzymes to diffuse and reach cleavable sites. Condensation may make condensed gel virtually immune to bacterial enzymatic attack. This is another remarkably interesting open question as it relates to mechanisms whereby polymers networks may become recalcitrant in SW.

4. Field Verification of the Presence of MG over the SW Column

DOM is the main nutritional substrate to bacteria and higher trophic levels in SW. However, DOM is largely composed of small recalcitrant moieties not available to bacteria [9,42]. A standing enigma in marine carbon cycling is why the main source of bacterial nutrients—and thereby the nutrient source for the rest of the marine food web—is largely constrained to large size molecular components of DOM [9,44]. The finding that MBP can self assemble forming MGs that contain four orders of magnitude more nutrients than DOM diluted in SW [2]—that can be readily colonized by microorganisms—provides an interesting alternative to explain this paradox. However, validation of this hypothesis requires to first demonstrate that gels like those found to self assemble in the laboratory are present in SW; and second, that bacteria do indeed feed preferentially in MGs.

To address the first question, we developed and validated a flow cytometry (FC) assay to count and directly measure the concentration of self-assembled microgels (SAG) in native unfiltered seawater [55].

Experiments to measure the presence and concentration of MGs in SW were conducted in samples collected at Hood Canal (47°50′ N; 122°38′ W); Admiral Inlet (48°10′ N; 122°38′ W); Strait of Juan de Fuca (48°21′ N; 124°22′ W); BATS station (30°50′ N; 64°10′ W); and particularly during our cruise aboard the R/V Kilo Moana to the Hawaii Ocean Time Series at ALOHA station (22°26′ N; 158°5′ W).

The presence of MG over the whole water column—from surface to 4000 m deep—was investigated in SW samples labeled with Chlortetracycline (CTC) and detected and counted using flow cytometry [55] and fluorescent quenching assays [56]. Notice that CTC labels Ca bound to biopolymers but does not crosslink or affect in any way the supramolecular structure of gels.

Outcomes of these studies were reported to the Faraday Discussions of Royal Society of Chemistry in 2008, [55]. Results confirm that particles with similar features to those found in self-assembled gels in the laboratory are indeed present in the ocean. i.e., they can undergo a phase transition and readily disassembly following Ca chelation of SW. As expected, the concentration of MGs over the water column follows very closely the typical concentration of DOM from which MGs are formed (Figure 21). Physical hydrogels like MG contain only 1–2% of solid and their density is largely determined by their water content. Thus, it is very unlikely that the MGs we found in the dark ocean may be settling down from the surface. Results of these studies imply that assembly must be taking place down the water column, a feature consistent with the profile of DOM concentration that is the feedstock for gel formation.

Surprisingly, even within the constraints of limited sampling, the range of concentration of MG found in this study agrees with both the magnitudes of global mass and global volume of MG predicted from a 10% thermodynamic yield of DOM assembly measured in the laboratory [2].

Indeed, flow cytometry measurements in Figure 21 show that MG in the 5 ± 3 μm size—with a corresponding average volume of ~7×10^{-14} L—reach concentrations ~10^8 MGs $\times$ L^{-1} of SW, yielding a MG volume to SW volume of the ratio of 10^{-16} L of gel per litter of SW. These figures, when scaled to the ~10^{21} L global volume of seawater in the planet [57], give a global volume of ~10^{15} L of MG. On the other hand, measurements in gels resulting from MBP self assemble in the laboratory indicate that ~70 Gt (7×10^{16} g) of reduced organic carbon is predicted to be present in the ocean forming MG [2]. Given that an average hydrogel contains only ~1% *w/v* of solid, the estimated global volume of MG could reach ~7×10^{15} L. Thus, 7×10^{15} L of MG calculated from the thermodynamic yield of DOM assembly in the laboratory [2] is remarkably similar to the 10^{15} MG $\times$ L^{-1} volume inferred from measurements of MGs in the field [55].

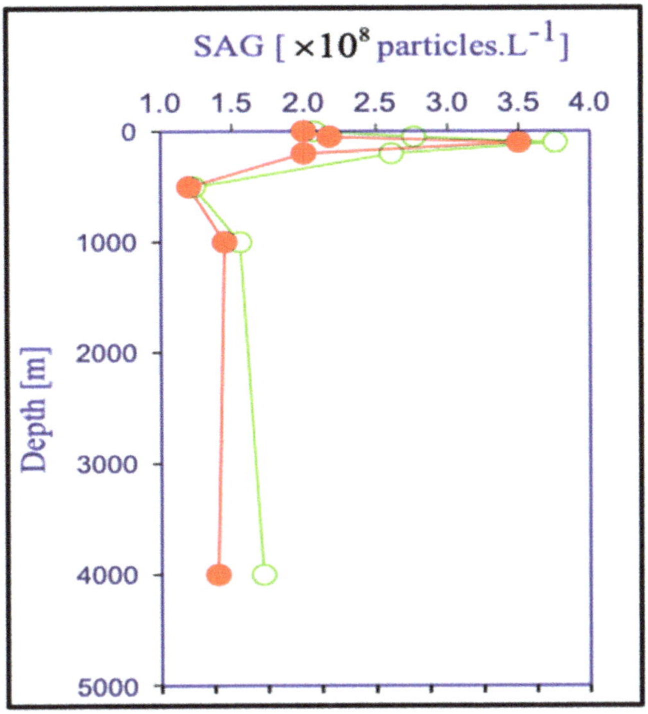

Figure 21. Typical results of MGs detection by flow cytometry—in closed red circles—and fluorescent quenching method—in open green circles. MG was fluorescently tagged with Chlortetracycline (CTC) that labels bound Calcium present in MG. Notice that the concentration of MG follows very closely the concentration of DOM MBP, which is the feedstock for self-assembly. Notice also that while CTC colorimetric method allows flow cytometry to count MG, and to image gels by fluorescence microscopy, it does not allow measurements of reduced organic carbon in MG.

A corresponding similarity can be found concerning the global carbon mass present in MGs. At an average gel density of $1 \times cm^{-3}$ and considering that gels contain ~1% *w/v* of solid, imply that a global volume of gel of ~10^{15} L must contain ~10^{16} g of solid, mostly organic material with a carbon content not far from the 7×10^{16} g of reduced organic carbon measurements in gels assembled in the laboratory [2].

Although there were variations of gel counting in different locations and depth, it is remarkable that figures of global gel volume and gel carbon content seems to be analogous in orders of the magnitude scale. Perhaps more significant is that a similar range of mass transfer can be inferred from Santschi's pioneering studies of DOM/POM (particulate organic matter) transformation using radioactive tracers [58,59]. Although the rates at which bioactive elements pass through the MG pool are unknown, the similar day-to-week time scales of MG formation and ^{223}Th pumping from colloidal to particulate size [58] suggest that the corresponding fluxes could be very large indeed.

Field results discussed above are still preliminary and need to be confirmed in broader field measurements of MG concentration in SW. However, the reported concentrations of MG [55] are remarkably consistent with thoroughly verified yields of self-assembly measured in the laboratory [2] giving further support to the notion that a global stock of ~70 Gt (7×10^{15} g) of reduced organic carbon is most likely present as marine gel reduced organic carbon (MGOC) in the world ocean [2,55]. The huge magnitude of this budget has far-reaching significance for marine carbon cycling [3,16]. A reversible dissolved MBP↔MG assembly process represents a critical mechanism whereby MBPs in DOC are transferred from a low (~1 mg × L^{-1}) concentration dissolved in SW—where it is relatively inaccessible to microorganisms—to a porous gel stock made of loosely interconnected polymers containing a solid/water ratio of ~1% yielding a 10^4 increase in local DOM concentration, that is readily available to bacterial enzymes and metabolization. The most significant implication is that DOM biopolymer self-assembly into MGs drives a continuous DOC↔MGOC carbon flux with a corresponding massive nutrient-rich pool of MGOC that enters the marine biological carbon processing escalator. This is a major shunt of mass transfer into the marine carbon cycling process. It challenges some conventional paradigms regarding processes linking the microbial loop and biological pump to the rest of the biosphere and the geosphere. A dynamic equilibrium between free and assembled DOC occurring over the whole water column produces micron-dimension gel patchiness that may help explain carbon turnover, particularly in the dark ocean. It may drive a massive flux of locally produced—not transported down—MGOC into the microbial loop, with ramifications that scale to global cycles of marine bioactive elements [3,16].

5. Bacterial Colonization of Marine Gels

5.1. Preliminary Field Verification of Bacterial Colonization of Marine Gel

MGs contain four orders of magnitude more microbial nutrient concentration than DOM. The next question to address is whether bacteria feed preferentially in these nutrient-rich networks as compared to the huge stock of dissolved organic moieties present in SW. A significant implication of this question is that if MGOC is an enriched source of microbial nutrients the distribution of bacteria in SW should not be isotropic but discrete. Bacterial concentration in SW should follow a Markovian rather than a Gaussian spatial distribution. Pioneering observations by Azam [4] confirm that bacteria are indeed present in random patchy clusters of high bacterial concentration [60]. However, whether these patches correspond to MG and how much more bacteria are found in MG as compared to SW has not been precisely established. There are multiple reports of MB found attached to particles or transparent exopolymer particles [60–62]. However, quantitative evaluation of bacteria lodged inside MG has not been systematically explored. We approached these questions by simply defining a partition coefficient of bacterial colonization, by measuring the ratio of MB concentration per volume of gel as compared to the concentration of bacteria per equivalent volume of SW. There are severe methodological limitations to test this idea though. MG are highly porous loosely tangled networks that bacteria can readily penetrate, but they are not transparent. To precisely measure the number of bacteria lodged inside a cloudy environment it is necessary to use confocal optics and thin section serial 3D optical tomography of MGs (Figures 22–24).

Labeling MB with BackLight Red^{RM}, a commercially available supravital fluorescent probe, and MG with chlortetracycline (CTC) allows the use of fluorescence confocal microscopy to readily image MG, bacteria lodged in MG, and free bacteria outside MG. By filtering a known volume of SW containing both colonized gels and free bacteria, the application of confocal microscopy allows to image and count free bacteria and gels landed on the filter. Confocal fluorescence microscopy allows to further perform multiple thin sections of the infected gels retained on the filter to count the number of MB per section (Figures 22 and 23).

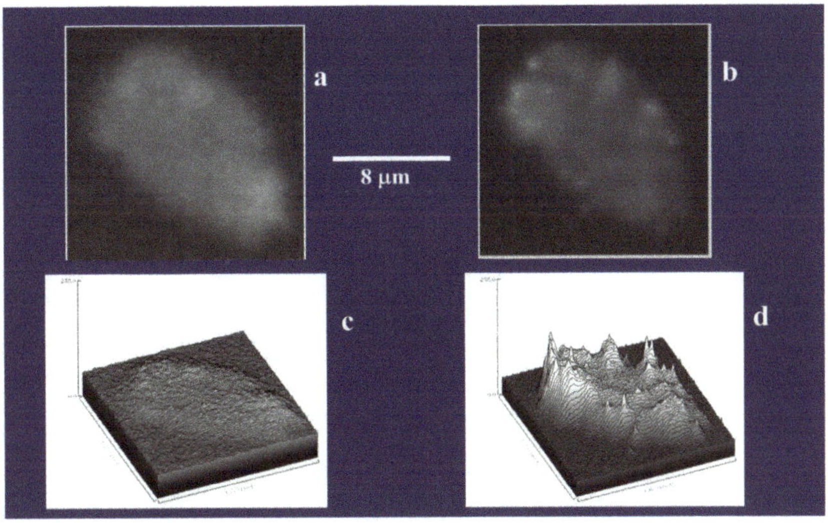

Figure 22. These images illustrate the advantage of using computer-processed thin sections of confocal imaging to eliminate out-of-focus image noise and precisely count the number of MB per section of gel. In here MG were labeled with CTC (λem = 560 nm), and MB labeled with BacLigh-Red^{TM} (λem = 644 nm). Insert (**a**) is a raw fluorescence image across the whole gel labeled with both CTC and BacLigh-Red^{TM}. Notice that that despite double fluoresces labeling, imaging across the whole MG does not allow to distinguish bacteria. Insert (**b**) shows a 300 nm thin gel section where bacteria can be readily observed. The corresponding 3D plots of contrast ratios of BacLigh-Red^{TM} are in (**c**) and (**d**). Notice that the thin section in (**d**) shows both the perimeter of the MG depicted by CTC fluorescence and the number of MB which can be identified by the sharp contrast peaks of BacLight fluorescence. It allows serial section tomography software, developed in our laboratory [29], to readily quantitate bacteria lodged inside MGs, avoiding double-counting by defining the specific x-y coordinates of every peak in each thin section.

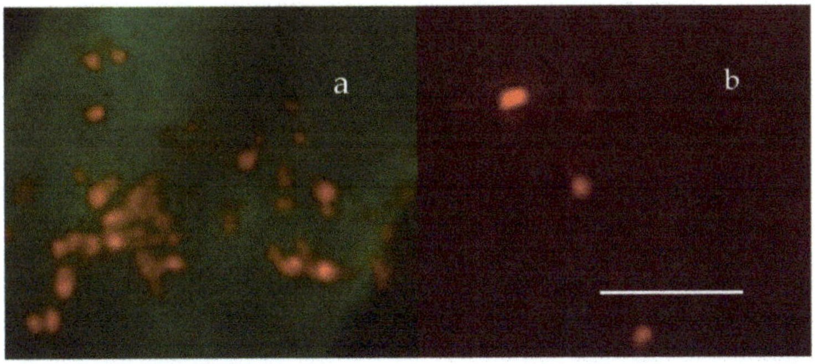

Figure 23. Panel (**a**) shows ~20 MB—labeled with BacLigh-Red^{TM}, found inside an ~100 μm^2 by 300 nm deep thin section of a MG, labeled with CTC. Panel (**b**) illustrates an equivalent field in the same preparation showing free MB—outside the MG—retained on top of the 0.2 μm pore filter. The bar = 10 μm.

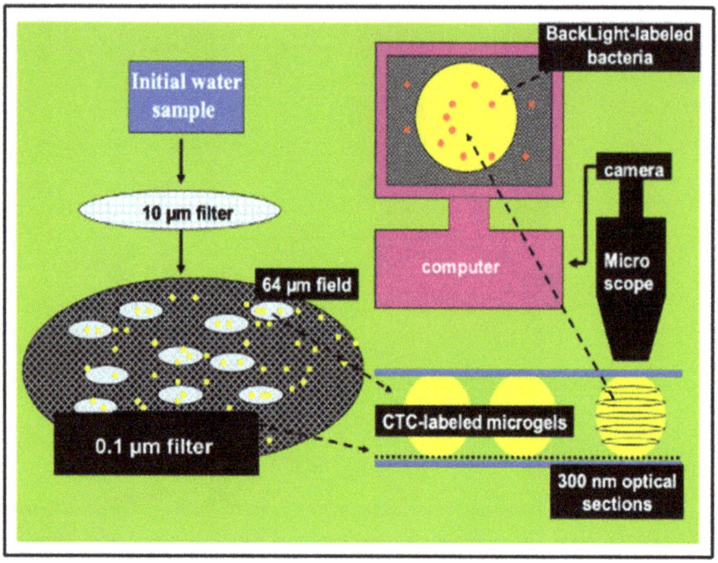

Figure 24. This is an illustration of the method we implemented to count and measure the volume of MG, and count free MB and MB lodged inside MGs. Aliquots of 100 mL of SW are inoculated with: 1 mM sodium azide to arrest bacterial growth; 10 mM BacLigh-Red^{TM} (λex = 581 nm, λem = 644 nm) to label bacteria; and 100 µM chlortetracycline (CTC) (λex = 374 nm, λem = 560 nm) to label MG (Figure 12). SW is first pre-filtered at 15 µm to retain larger particulate material. The filtrate is then filtered at 10 cm Hg vacuum pressure through a 0.1 µm black Nucleopore® filter that retains both MB and MGs. The black filter is then set between two coverslips suspended by 4 µm microspheres to avoid compressing the MGs. This preparation is then positioned in a Nikon inverted fluorescence microscope with MCR-600 BioRad series laser scanning confocal imaging to perform thin sections tomography of MGs. Counting of both MGs and free bacteria found on top of the filter—outside MG—is conducted in randomly selected fields of ~65 µm diameter. The average ± SEM counts of free bacteria and gels found in 20 fields are scaled to the total diameter of the filter to report the total of free bacteria and MGs found in the 100 mL aliquot of SW that passed through the filter. Tomographic thin sections of ~300 nm of MGs are imaged at 2000× magnification and serially sampled at 1 µm spacing. MG volume is reconstructed by the computer based on the volume of thin sections over the whole serial stack. Bacterial count is performed by tomographic software based on the number of peaks found in 3D plots of contrast ratios of BacLigh-Red^{TM} emission (See Figure 22). In these assays, concentration MB in MG is based on averages ± SEM of counts in 3 to 5 tomographic sections per MG in randomly selected MGs found on five filters per sample.

Preliminary results illustrated in Figures 25–27 were presented at the 2007 ASLO meeting. They correspond to seawater collected at 10, 50, and 100 m transect in the San Juan Channel (48.45° N, 122.96° W, on 11/04/2006, samples were immediately inoculated with 1 mM sodium azide to stop the bacterial activity. Five 100 mL aliquots of each sample were then processed according to the protocol described in Figure 24.

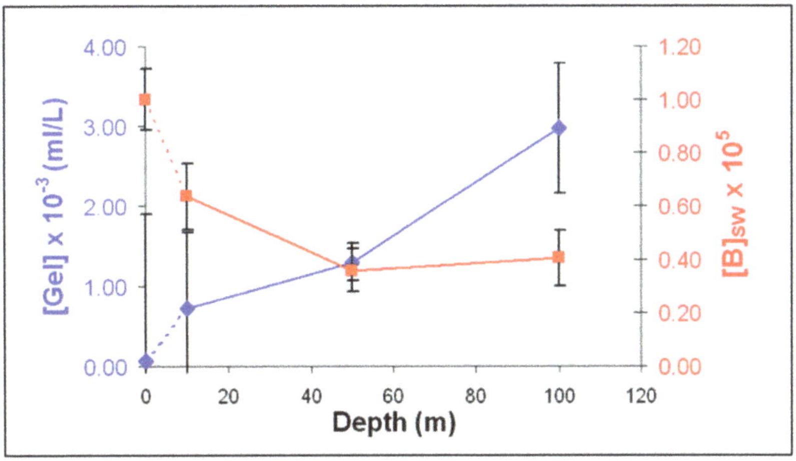

Figure 25. Images of four fields captured at 1200× magnification, covering 60–80 μm^2 each were randomly selected and recorded in each of five filters containing fluorescently labeled gels and bacteria found in the surface, 10, 50, and 100 m deep seawater samples. These images yielded 105 data sets reporting the count of MG and free MB found in each of the fields. These numbers were then averaged and scaled to the 78 mm^2 areas of the filter to render the concentration of bacteria in seawater ([MB]SW = ΣBSW × mL^{-1} SW) and the SW gels concentration ([MG] = ΣMG × mL^{-1} SW) found in the 100 mL of seawater that passed through each filter. The volume of each MG (VMG) was calculated from the area and thickness of each serial optical section and the number of sections in the stack.

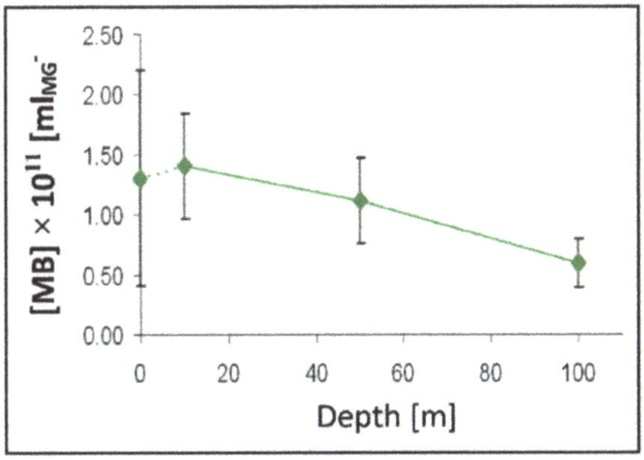

Figure 26. The concentration of $[MB]_{MG}$ lodged inside MG ranged from 0.6 to 1.4 × 10^{11} MB × mL^{-1} SW. It is more than five orders of magnitude higher than the concentration of free [MB] in SW. Points are the average ± SD. of 72 sections in 12 microgels collected in each of two samples per depth.

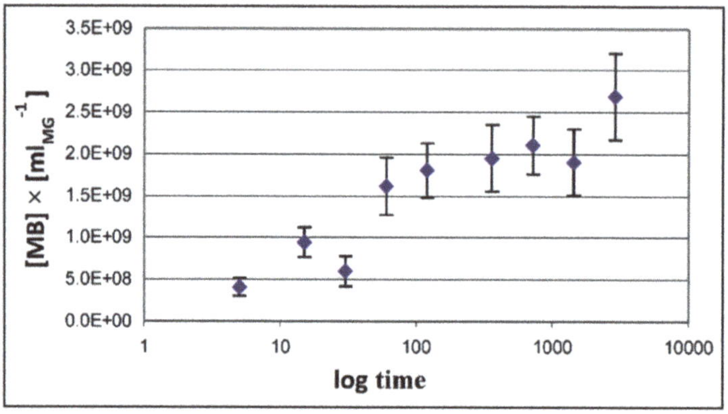

Figure 27. Nine 100 mL samples of 0.2 µm-filtered seawater were incubated for 144 h. to produce self assemble gels. Each sample was then inoculated with bacteria isolated by filtration from seawater collected from the dock of Friday Harbor Laboratories and incubated at 10 °C. MB growth was then sequentially arrested in samples 1 through 9 every 80 min by exposure to 1 mM NaN3, and fluorescently tagged with CTC to label MG and BacLight Red to label MB. Each sample was then immediately filtered using 0.1-µm black Nucleopore filters. Filters were mounted in a Nikon fluorescence confocal microscope. Images of four fields captured at 1200× magnification, covering 60–80 µm^2 each were randomly selected and recorded from each filter containing fluorescently labeled gels and bacteria found on the surface of the filter. MB lodged inside gels were counted in 300-nm confocal sections of MG. The points are the average ± SD of 256 sections in 16 microgels collected in each filter. Although the concentration of bacteria in seawater remained at ~2 × 10^6–9.2 × 10^6 bacteria × mL^{-1}, within few minutes the concentration of bacteria inside microgels equilibrated with the bacterial concentration in seawater and then increased exponentially with time from ~2.8 × 10^6 bacteria × mL^{-1} of gel to ~2.7 × 10^9 bacteria × mL^{-1} of gel. Points are the average ± s.d. of 36 sections in 12 microgels collected in each of the nine filters corresponding to successive 80 min arrest of bacterial activity.

The average MG volume expressed in mL (V_{MG}) can then be inferred from measurements performed in 75 gels. From the product of gel concentration [MG] SW and $\overline{V}$MG we then obtained the volume ratio of MG to SW in VMG × L^{-1} SW in mL MG × mL^{-1} SW. These data provided a robust statistical profile of the volume of MG × L^{-1} SW in a litter of SW, which ranges from 1 to 4 µL MG × L^{-1} SW. Similarly, free MB found on the surface of the filter (Figure 23b) ranged from 0.4 to 1.2 × 10^5 × mL^{-1} SW. Each point corresponds to the average ± SD of 105 measurements.

5.2. Dynamics of Bacterial Colonization of Marine Gels

In the experiment illustrated in Figure 27, MG self-assembled from 0.2 µm filtered DOM were inoculated with MB from raw SW sampled in Friday Harbor Bay. While over the whole incubation MB count outside gels remained unchanged at ~2 × 10^6 bacteria mL^{-1} of SW, bacterial colonization of MG grew exponentially to reach ~2.7 × 10^9 bacteria mL^{-1} of gel by the end of 12 h incubation.

Although these preliminary observations did not receive much attention at the ocean science meeting in 2007, they offer striking evidence that MGOC might be the primary substrate for MB. Bacteria lodging inside MGs is like Hansel and Gretel in the cookie house, where their enzymes can cleave a largely immobilized substrate present at high concentrations in the gel matrix. The model illustrated in Figure 16 predicts that as MGOC is consumed new MGOC is formed, securing a supply of nutrients whose original feedstock is the huge mass of reduced carbon present in the DOC pool.

Data in Figure 25 indicate that in 1 L of SW there are ~10^{-6} L of gel. In 10^{-6} L of gel—according to data in Figure 24—there are ~10^8 bacteria. Published figures of bacterial concentration in the ocean report ~10^9 bacteria per litter of SW. Thus, according to our findings, the actual concentration of marine bacteria—including bacteria lodged in gel plus free bacteria—must probably be in the order of ~10^{17} bacteria per liter of seawater. Published figures might likely be severely underestimating the concentration of bacteria present in the ocean. Scaling ~$10^{17} \times L^{-1}$ of bacteria in SW to the ~10^{21} L of SW in our planet [57], suggests that 10^{38} MB are likely to be present in the world ocean. This figure is eight orders of magnitude larger than previous estimates of the global number of bacteria on the planet [62].

Considering that bacteria are ~0.5 μm^3 and given an estimated to contain ~10% solid, implies that a bacterium must have a mass of ~10^{-8} gr of organic solid. This means that that a global biomass of ~10^{30} g of organic material may probably be present and transported up the food chain by MB in the world ocean. This is 8 orders of magnitude higher than the ~10^{18} gr of DOM present in the ocean, it implies that the carbon flux passing through the MGOC must be very high, or that the flux of organic carbon transported by marine bacteria (MBOC) to higher trophic labels may be significantly lower than the DOC↔MGOC carbon flux.

It is worth pointing to another singular agreement. Namely, these studies—counting MG by fluorescence confocal microscopy—a very different protocol of the flow cytometry studies described in Section 4—give a MG/SW ratio of 10^{-6} L of gel per litter of SW, which is the same figure obtained by confocal microscopy.

While the data of these studies are robust, the result reported here are limited to a single instance in one geographical location and within a narrow span of depth sampling. They need to be verified at a broader scale. Nonetheless, if these figures hold, it will require a thorough revision of the methods and the strategies presently used to study MG and the MG-microbiology of the ocean.

6. The Gel Pathway to Carbon Flux in the Ocean

There are two old overarching and highly controversial enigmas in the complex path of carbon flux in the world ocean. One pertains to what are the molecular or supramolecular features that make MBP susceptible to bacterial consumption [63]; the other is the converse, namely, what features condemn a DOM moiety to join the recalcitrant nonreactive pool of old—some of it thousands of years old—DOM molecules present in SW [9,43].

Amon and Benner's size-reactivity hypothesis proposes that the bioreactivity of natural organic matter decreases along a continuum of molecular size [63]. The model does not explain why bacteria discriminate based on size when choosing their substrate. There are three confluent lines of explanation for the mechanisms underlying the size-reactivity idea. All point to the hypothesis that bacteria don't feed in free polymers, no matter their size, but on assembled DOM biopolymers present in the matrix of MGs. One line is derived from polymer networks theory. Namely, the probability of polymer self-assembly increases with the second power of polymer size [11–15,64], i.e., HMW DOM (high molecular weight DOM) has a quadratic advantage over LMW (low molecular weight) species when interacting to form networks and gels. This gives HMW DOM an enormous advantage, over smaller moieties, to self-assemble and join the reactive MGOC pool. The other two lines come from our work. First and foremost, the demonstration that MBP in DOM does indeed self assembles [2]. The point of interest, in this case, is that—as shown in Figure 13—self assemble occurs in times very similar to the incubation times of Benner's experiments, which are the base of the size-reactivity hypothesis [63]. Chin et al. data [2] show that within 24 h, DOM can self assemble forming MG of about 2 μm hydrodynamic diameter; and by 48 h, the assembly has reach equilibrium forming gels of about 4 μm diameter (Figure 13). Benner's results show that both bacterial cell count and Leucine incorporation peak after 48 h of incubation. The second line of evidence that can help to explain Benner's results is in Figure 27. It shows that in the presence of gels, bacterial numbers grow exponentially. Within 16 h MB count in MG can reach ~2×10^9 bacteria per ml of gel. These results are certainly not proof that in Benner's experiments bacteria were probably feeding on gels, although, it is a very persuasive correspondence that invites experimental tests.

Nevertheless, the points raised here are not objections to the size-reactivity hypothesis. Benner's hypothesis is fundamentally right. However, the reason why HMW DOM is more reactive is most likely because this fraction of DOM is made from larger, highly charged polyelectrolytes and amphiphiles biopolymers that can readily for MGs. i.e., macromolecules in the HMW pool obey a physical law that makes them self assemble and form MGOC that is probably the actual reactive substrate for marine bacteria. In short, polymer theory sets a very strict rule of selection of the species that can or cannot self-assemble and are correspondently reactive or left on the recalcitrant non-reactive pool.

A critical corollary of de Gennes and Edwards laws—on which is based the gel pathway hypothesis for marine carbon flux proposed here—is that short polymers should fail to assemble. Two lines of experimental evidence indicate that marine biopolymers are no exception to the rules that govern polymer networks dynamics [11–15,64]. Cleavage of DOM by two independent methods confirms this corollary.

Ultraviolet radiation, (UV) routinely used in industrial polymer cracking processing, provides a convenient way to test the principle that downgrading the polydispersity of DOM stock should inhibit self-assembly. The results illustrated in Figure 28 show that depending on the length of time that DOM has been exposed to UV-B, their self-assembly kinetics slow down, and the equilibrium size of assembled gels decreases progressively. After 12 h of UV-cracking chain size must decrease sufficiently as to prevent self-assembly and gel formation. To address previous and recent unwarranted criticisms [65,66], notice that in these experiments the radiated species were not gels, but freshly 0.2 μm-filtered DOM polymers that were subsequently allowed to self assemble, in absence of UV light.

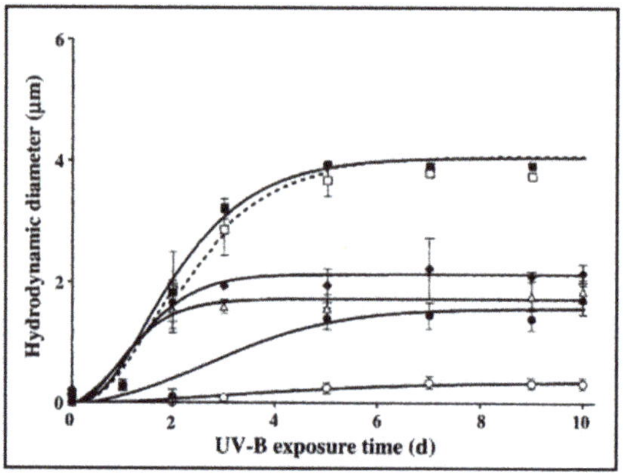

Figure 28. In these experiments, six aliquots of 0.2 μm-filtered SW were UV irradiated for a progressively longer time, and then DOM was set to self assemble under continuous monitoring of particle size Dynamic Laser Scattering. Control not irradiated DOM (filled squares) or irradiated for 24 h at 10 W m^2 with UV-A light (λ = 230–400 nm (open squares). Samples of DOM irradiated at 500 mW m^2 with UV-B (λ = 280–315 nm) for 30 min (filled diamonds) 1 h (open triangles) 6 h (filled circles) or 12 h (open circles). Results reveal that the DOM polymer assembly in non-irradiated or samples by UV-A follows the typical second-order kinetics described in Chin et al. [2]. However, as UV polymer cracking of DOM polymers increases with time of exposure to UV-B, DOM assembly and gel formation slow down, equilibrium size of the microgels size decreases, and finally, assembly virtually fails in DOM exposed to UVB for 12 h. Data points correspond to the mean ± s.d of 30 DSL measurements.

Another instance to validate de Gennes's [11] predictions on self-assembly of MBP, is that bacterial enzymatic cleavage of DOM should inhibit self-assembly. Experiments to test the effect of bacterial enzymatic cleavage on DOM self-assembly consisted of exposing five 100 mL aliquots of 0.2 μm-filtered SW to bacterial inoculum for 1 to 19 days. After completing their corresponding period of bacterial exposure each sample was treated with 1 mM NaN_3 to arrest bacterial activity, filtered by 0.2 μm pore filter, and set in the laser spectrometer to monitor DOM self-assembly and gel formation.

As shown above (Figure 29), as DOM is progressively exposed, bacterial enzymatic cleave, the assembly kinetics slow down, the equilibrium size of the assembled microgels decreases, and—like in the case of UV cracking—after 19 days of bacterial exposure, cleaved DOM polymers fail to self-assemble (Figure 29).

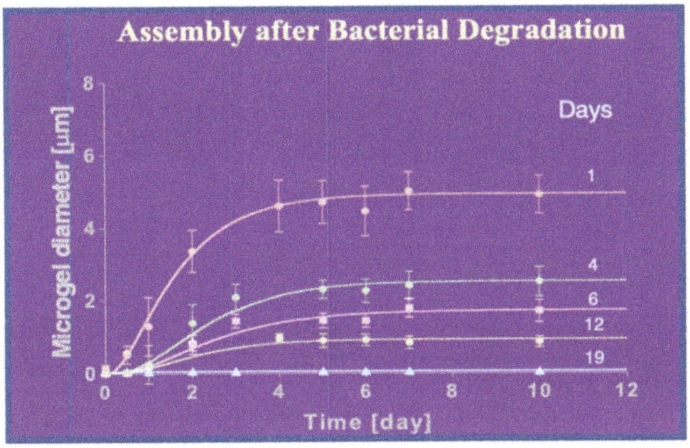

Figure 29. Bacterial cleavage of DOM results in progressive compromise of MBP self-assembly. The protocol for these assays is described above. Data points correspond to the mean ± S.D of 18 DSL results.

What makes DOM biopolymers recalcitrant non-reactive has been for a long time the subject of much controversy in geochemistry [9,42]. The evidence presented here allows us to suggest a very simple hypothesis. Namely: the bioreactivity of DOM biopolymers depends largely on their ability to self-assemble forming MGs that bacteria can colonize, cleave, and metabolize. DOM assembly is governed by fundamental principles of polymer networks that enforce a strict quadratic size selection that determines what biopolymers can self assemble forming gels and be reactive and what others cannot self assemble and are condemned to the recalcitrant pool. The relevant functionalities that save DOM biopolymers from going to the recalcitrant waste basket are their size, charge (Z potential), and hydrophobicity, all features that strongly influence self-assembly.

This size-assembly-bioreactivity hypothesis presented here does not rule out the production and presence in SW of species that are or have become non-reactive by specific molecular modifications that make them resistant to bacterial enzymes (Figure 1).

The last illustration in Figure 30 was drawn by my friend John Hedges. It was included in the application we submitted to the NSF Engineering Directorate that supported most of the studies reported here. At the time, in 2000, the idea illustrated in Figures 1 and 30 was just a guess coming out of Hedges-Verdugo brainstorming and reciprocal teaching of Oceanography 101-Polymer Physics 101. Today, there is enough robust preliminary evidence to consider that this biopolymer-self-assembly-bioreactivity hypothesis has a good chance to be correct and to use it as a convenient compass to guide future inquiry on the multiple pending riddles on marine gels and global carbon cycling.

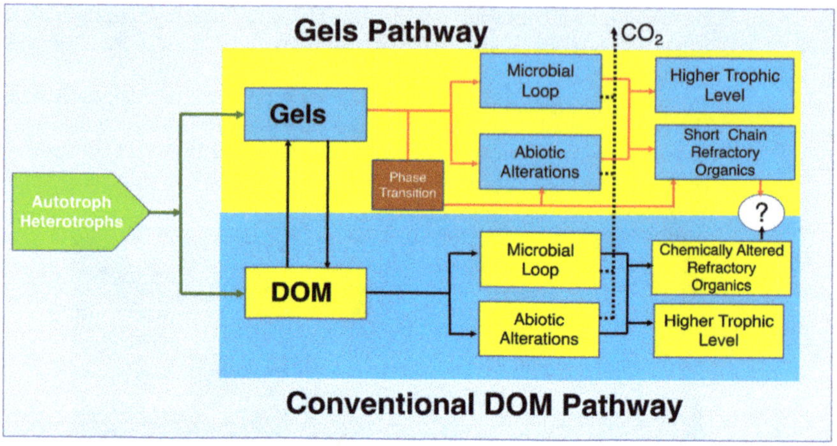

Figure 30. Molecular reduced-organic-carbon-carriers probably have multiple ways to move up through the marine consumer's conveyor. Two hypothetical pathways are illustrated here; one that postulates that DOM becomes directly available to marine bacteria primarily based on their molecular weight. The other proposes that bacteria feed primarily in marine gels made of large HMW moieties. According to the DOM hypothesis, the pass for moieties to become refractory is by potential chemical alterations that make these molecules resistant to bacterial enzymes. The Gel pathway hypothesis postulate that, while there are chemically altered species in SW, refractory molecules are simply those too large to be metabolized by bacteria and too small to self assemble. Implicit in this statement is the notion that marine gels are necessary and probably sufficient to drive carbon flux dynamics in the ocean.

7. Intellectual Merit

The "DOC-MGOC mass transfer hypothesis" is based on laboratory observations that (1) DOM remains in reversible assemble↔dispersion equilibrium forming microscopic marine gels (2) The thermodynamic yield of DOM self-assembly amounts to ~70 billion tons of nutrient-rich organic matter. (3) Marine gels are present over the whole water column with an estimated global gel volume of ~7×10^{15} L. (4). Reversible dissolved↔assembly process represents a mechanism whereby DOM MBPs move from being dissolved at low (~1 mg L^{-1}) concentration to forming a porous gel matrix containing a solid/water ratio of ~1%. It means a 10^4 increase concentration found in MGOC, where the substrate is virtually immobilized readily available to bacterial enzymes cleavage and metabolization. Hence, gel formation drives a continuous DOM flux that produces a huge nutrient-rich pool of organic substrate; it is a major shunt of mass transfer into the marine carbon cycling process. (5) That short residues, including those resulting from bacterial enzyme cleavage, fail to form gels, and are likely to comprise the bulk of the recalcitrant stock present in the ocean. (6) Marine bacteria reach ~1.4×10^{14} MB per litter of gel, this is 5 orders of magnitude higher than the ~10^9 bacteria per litter of SW. Considering that there are only ~10^{-6} L of gel in a litter of SW, the actual concentration of marine bacteria—including bacteria lodged in gel plus free bacteria—may probably reach ~10^{17} bacteria per liter of seawater. Thus, it is likely that most published reports might be severely underestimating the concentration of bacteria present in the ocean, which implies that about one-half of the marine bacterial population may have probably been largely ignored in the past.

These observations lend strong support to the hypothesis that DOM self-assembly and microgel formation may afford unique mechanisms and pathways for the flux of dissolved organic carbon substrates to bacteria and higher trophic levels, eventually affecting global elemental cycles and atmospheric CO_2 dynamics. They offer a simple, polymer theory-based, explanation for the nature and mechanisms of production of the recalcitrant stock found in the world ocean.

8. Conclusions and Homework

It often takes an outsider to find that the king needs dressing up. In this case, it seems Neptune is missing his crown and has no fisherman's spear.

(1) Detection and quantitation of marine gel and particularly carbon present in gels need to improve. See point 1 in Appendix A.
(2) There is a need to validate the preliminary results presented here on bacterial count lodged in MG. To do so, it is critical to find a simpler method to quantitate bacteria found inside gels. Optical tomography is a robust and elegant method to image and count bacteria, but, despite computer automation, it is a painfully time-consuming technique—check point 2 in Appendix A.
(3) To gain an understanding of self assemble turnover in the dark ocean it is important to find out the effect of pressure in DOM self-assembly and particularly on the eventual phase transition of MG.
(4) The chemical composition of self-assembled gels remains largely inferred from polymer theory predictions and indirect fluorescent probes labeling. It needs to be systematically explored.
(5) The cell biology of phytoplankton needs to be better investigated. Phytoplankton are complex secretory cells whose most central function—critical to carbon flux in the ocean—requires a clear understanding of their cell physiology and biophysics, away from untested dogmas and phenomenological descriptions. The finding that phytoplankton function as secretory cells opens a powerful conceptual paradigm of cell secretion that can serve as a guide to further explore the cellular mechanisms that control phytoplankton MBP production.
(6) The chemistry of MG remains as the chemistry of blood coagulation in the early nineties: rich in phenomenology and hypothesis, yet largely orphan of mechanistic science. Seawater is a complex polymer soup, where chemical interactions involving electron donor-acceptor transactions are not the most common currency. The macromolecular thermodynamics of marine biopolymers rests largely on polymer physics territory. Macromolecular interactions without chemical exchange of electrons are likely to be the rule in the Neptune kingdom. Hence, awareness of the laws that govern the physics of MBP is critical to design strategies aimed at understanding and predicting the complex macromolecular interactions that lead to MG formation.
(7) The intended goal of this paper is the hope that the next generation of oceanographers get a thorough formal training in polymer physics. After we finished writing the application that funded the studies described here, John Hedges was committed to—at his return from sabbatical leave in Germany—persuade colleagues in our University of Washington departments of Physics and Oceanography to design a pilot class of polymer theory for Oceanography undergraduates. Fate prevented this dream. Polymer Physics and Polymer Networks Theory is still missing in most Oceanography curriculums. It's time for someone to take the torch.

Funding: This research was funded by Grant 0120579 from the Biocomplexity Program of the National Science Foundation Div. Bioengineering and Environmental Systems to P.V.PI, and JIH coPI.

Institutional Review Board Statement: Not applicable.

Informed Consent Statement: Not applicable.

Acknowledgments: Finally, this research would have not been possible without the collaboration of John Hedges and his wide-angle and deep perspective of marine geochemistry; or the seasoned advice of my friends Toyo Tanaka, Sam Edwards, Karel Dušek, and Peter Santschi; and without the support of Dennis Willows, former Director of Friday Harbor Labs; and the collaboration of Jekaterina Zyuzin, John Tuthill, Anthony Rodriguez, Andrew Moon, Chung Jong Yu, Tsveta Krumova, Jamila Harris, Jordan Stead, Maria Jose Bravo, Kelly Kettleson, Meena Padha, Thien Nguyen, Wei-Chun Chin, and Yongxue Ding; and former research fellows Ivan Quezada and Monica Orellana. Finally, I'm equally thankful for the outstanding support of our technical assistant and cat-herding boss

Michelle Herko, who kept everyone at the bench, supplies on time, and our instruments running and well serviced.

Conflicts of Interest: The author declares no conflict of interest.

Note: The material reported here is a summary review of research on marine biopolymer dynamics conducted in our laboratory. It is intended to illustrate the application of polymer physics theory to investigate the role of gels on marine carbon cycling. The manuscript does not follow the standard format of research communication, nor the format of a literature review. References are restricted to the subjects discussed in the text. For a thorough review of the author on marine gels, readers are invited to consult [16]. There are many definitions of gels, and it is broadly acknowledged that gels are hard to specifically define and classify. For a thorough discussion on polymer gels, the reader may check Refs. [14,15,64,67].

Appendix A

We are at the head of a trail to crack some of the most fundamental Neptune riddles on marine carbon cycling. Here are a couple of simple protocols to follow the path:

1. We now know the molecular mechanisms whereby MBP forms gels. The kinetics and thermodynamic yield of these process has been reliably measured and reported [2]. These gels contain a continually renewed ~70 billion tons of reduced organic carbon and they are likely to be indeed present over the whole water column [55]. The flux of reduced carbon that transit through the marine gel phase—DOC↔MGOC carbon flux shunt—is critical to gauge the thermodynamics of marine mass transport between the DOC source stock and the microbial and recalcitrant sinks in the ocean. What follows is a simple dye-free and image-independent assay to determine the ratio of organic Carbon assembled in gels v/r organic Carbon present in non-assembled DOM in SW samples:

Set three aliquots A, B, and C of 250 mL in clean vials. A and B are filled with the SW analyte, C is filled with clean ASW (artificial seawater) as control. All three aliquots are passed through 10 μm filters. Vial A is refiltered at 0.2 μm pore. The filtrate, in this case, should contain DOM minus DOM assembled in MG, as MG and other particles are retained in the filter. Vial B is dialyzed for 40 h—in bags of 20 mL—against a tank containing 10 L of Ca-free ASW and then filtered at 0.2 μm pore size. Having lost all Ca crosslinkers, the filtrate of vial B must contain free DOM plus the DOM from disassembled gels. Vial C is a control and should undergo dialysis and filtration at 0.2 μm filter like B to account for organics eventually shed by dialysis bags and filters. Carbon in the three filtrates is then measured by high-temperature catalytic oxidation (HTCO). The background carbon shed by dialysis bags and filters is in filtrate C and must be subtracted from Carbon in A and B. Filtrate B correspond to total Carbon from assembled and non-assembled DOM. Filtrate A is Carbon corresponding to free DOM minus assembled (gels) DOM that in this case have been retained by the 0.2 μm filter. Carbon that is present in gels MGOC is the difference between filtrates B and A.

2. A similarly simple protocol can yield reliable SW counting of MB present in MG v/r free-MB (outside MG). By taking advantage of the fact that in absence of Ca MG readily disperses; dialysis in Ca-free ASW should free bacteria lodged in gels, allowing to readily count them. In this protocol two 100 mL aliquots A, and B, of SW are first filtered at a 10 μm pore filter. Both samples are then titrated with 1 mM NaN_3 to arrest bacterial activity and with 10 mM BackLight Red to label MB. Sample A is dialyzed for 40 h—in sterile dialysis bags of 20 mL—against a tank containing 10 L of Ca-free ASW, and then filtered at 0.2 μm pore size black filter. The filter is mounted in a microscope and bacteria can be readily counted by standard fluorescence microscopy. Sample B is not dialyzed but filtered using a 3 μm pore size filter that should retain most gels but not bacteria. The filtrate is then filtered using a 0.2 μm pore size black filter, that is mounted in the fluorescence microscope to proceed with bacterial counting. Counts from filtered A samples should reveal both

free and gels-lodged bacteria, while counts from filtered B samples should reveal free bacteria not lodged in gels. The difference between the two counts should give the number of bacteria present in gels. This protocol should be easy to adapt to conduct bacterial count by flow cytometry.

There are two build-in assumptions in this procedure though. One, that exposure to Ca-free SW will do not lysis bacteria, and second, that only a small fraction of MB will remain attached to dispersed MBP after gels are disassembled by loss of Ca crosslinking.

References

1. Nossal, R.; Chen, S.-H.; Lai, C.-C. Use of laser scattering for quantitative determinations of bacterial motility. *Opt. Commun.* **1971**, *4*, 35–39. [CrossRef]
2. Chin, W.-C.; Orellana, M.V.; Verdugo, P. Spontaneous assembly of marine dissolved organic matter into polymer gels. *Nat. Cell Biol.* **1998**, *391*, 568–572. [CrossRef]
3. Wells, M.L. A neglected dimension. *Nature* **1998**, *391*, 530–531. [CrossRef]
4. Azam, F. Oceanography: Microbial Control of Oceanic Carbon Flux: The Plot Thickens. *Science* **1998**, *280*, 694–696. [CrossRef]
5. Mague, T.H.; Friberg, E.; Hughes, D.J.; Morris, I. Extracellular release of carbon by marine phytoplankton; a physiological approach. *Limnol. Oceanogr.* **1980**, *25*, 262–279. [CrossRef]
6. Bjornsen, P.K. Phytoplankton exudation of organic matter: Why do healthy cells do it? *Limnol. Oceanog.* **1988**, *33*, 151–154. [CrossRef]
7. Alldredge, A.L.; Passow, U.; Logan, B.E. The abundance and significance of a class of large, transparent organic particles in the ocean. *Deep. Sea Res. Part I Oceanogr. Res. Pap.* **1993**, *40*, 1131–1140. [CrossRef]
8. Smoluchowski, M. Drei Vorträge über Diffusion, Brownsche Molekularbewegung und Koagulation von Kolloid teilchen. *Phys. Zeitschrift Physik.* **1916**, *17*, 557–571, 585–599.
9. Jiao, N.; Herndl, G.J.; Hansell, D.; Benner, R.; Kattner, G.; Wilhelm, S.; Kirchman, D.L.; Weinbauer, M.; Luo, T.; Chen, F.; et al. Microbial production of recalcitrant dissolved organic matter: Long-term carbon storage in the global ocean. *Nat. Rev. Genet.* **2010**, *8*, 593–599. [CrossRef] [PubMed]
10. Tanaka, T. Gels. *Sci. Am.* **1981**, *244*, 124–138. [CrossRef] [PubMed]
11. De Gennes, P.G.; Leger, L. Dynamics of Entangled Polymer Chains. *Annu. Rev. Phys. Chem.* **1982**, *33*, 49–61. [CrossRef]
12. Edwards, S.F.; Grant, J.W.V. The effect of entanglements on the viscosity of a polymer melt. *J. Phys. A Math. Nucl. Gen.* **1973**, *6*, 1186–1195. [CrossRef]
13. Guennet, J.M. *Thermoreversible Gelation of Polymers and Biopolymers*; Academic Press: London, UK, 1992.
14. Djabourov, M.; Nishinari, K.; Ross-Murphy, S.B. *Physical Gels from Biological and Synthetic Polymers*; Cambridge University Press: Cambridge, UK, 2013.
15. Horkay, F. Polyelectrolyte Gels: A Unique Class of Soft Materials. *Gels* **2021**, *7*, 102. [CrossRef]
16. Verdugo, P. Marine gels. *Ann. Rev. Marine. Sci.* **2012**, *4*, 375–400. [CrossRef] [PubMed]
17. Dušek, K.; Patterson, D. Transition in swollen polymer networks induced by intramolecular condensation. *J. Polym. Sci. Part A-2 Polym. Phys.* **1968**, *6*, 1209–1216. [CrossRef]
18. Tanaka, T. Collapse of Gels and the Critical Endpoint. *Phys. Rev. Lett.* **1978**, *40*, 820–823. [CrossRef]
19. Chin, W.C.; Orellana, M.V.; Quesada, I.; Verdugo, P. Secretion in unicellular marine phytoplankton: Demonstration of regu-lated exocytosis in *Phaeocyctis globosa*. *Plant Cell Physiol.* **2004**, *45*, 535–542. [CrossRef] [PubMed]
20. Quesada, I.; Chin, W.-C.; Verdugo, P. Mechanisms of signal transduction in photo-stimulated secretion in Phaeocystis globosa. *FEBS Lett.* **2006**, *580*, 2201–2206. [CrossRef]
21. Verdugo, P. Goblet cells and mucus secretion. *Annu. Rev. Physiol.* **1990**, *52*, 157–176. [CrossRef]
22. Tanaka, T. Phase transitions of gels. *ACS Symp. Ser.* **1992**, *480*, 1–21.
23. Thornton, D.C.D. Dissolved organic matter (DOM) release by phytoplankton in the contemporary and future ocean. *Eur. J. Phycol.* **2014**, *49*, 20–46. [CrossRef]
24. Aaronson, S. The synthesis of extracellular macromolecules and membranes by a population of the phytoflagellate Ochromo-nas danica. *Limnol. Oceanogr.* **1971**, *16*, L9. [CrossRef]
25. Palade, G. Intracellular aspects of the process of protein synthesis. *Science* **1975**, *189*, 347–358. [CrossRef] [PubMed]
26. Thiel, G.; Battey, N. Exocytosis in plants. *Plant Mol. Biol.* **1998**, *38*, 111–125. [CrossRef] [PubMed]
27. Tanaka, T.; Fillmore, D.J. Kinetics of swelling of gels. *J. Chem. Phys.* **1979**, *70*, 1214–1218. [CrossRef]
28. Verdugo, P. Hydration kinetics of exocytosed mucins in cultured secretory cells of the rabbit trachea: A new model. In *Ciba Foundation Symposium 109—Mucus and Mucosa*; Nugent, J., O'Connor, M., Eds.; Wiley: Hoboken, NJ, USA, 1984. [CrossRef]
29. Nguyen, T.; Chin, W.-C.; Verdugo, P. Role of Ca^{2+}/K^{+} ion exchange in intracellular storage and release of Ca^{2+}. *Nat. Cell Biol.* **1998**, *395*, 908–912. [CrossRef]
30. Katchalsky, A.; Lifson, S.; Exsenberg, H. Equation of swelling for polyelectrolyte gels. *J. Polym. Sci.* **1951**, *7*, 571–574. [CrossRef]
31. Donnan, F.G.; Guggenheim, E.A. Exact equation for thermodynamics of membrane equilibrium. *Z. Phys. Chem.* **1932**, *A162*, 346–360. [CrossRef]

32. Tam, P.Y.; Verdugo, P. Control of mucus hydration as a Donnan equilibrium process. *Nat. Cell Biol.* **1981**, *292*, 340–342. [CrossRef]
33. Aitken, L.M.; Verdugo, P. Donnan mechanism of mucin release and conditioning in goblet cells: The role of polyions. *J. Exp. Biol.* **1989**, *53*, 73–79.
34. Orellana, M.V.; Lessard, E.J.; Dycus, E.; Chin, W.-C.; Foy, M.S.; Verdugo, P. Tracing the source and fate of biopolymers in seawater: Application of an immunological technique. *Mar. Chem.* **2003**, *83*, 89–99. [CrossRef]
35. Passow, U.; Alldredge, A.L. A dye-binding assay for the spectrophotometric measurement of transparent exopolymer parti-cles. *Limnol. Oceanogr.* **1995**, *40*, 1326–1335. [CrossRef]
36. Engel, A.; Passow, U. Carbon, and nitrogen of transparent exopolymer particles (TEP) in relation to their Alcial Blue adsoption. *Mar. Ecol. Prog. Series* **2001**, *219*, 1–10. [CrossRef]
37. Thornton, D.C.; Fejes, E.M.; DiMarco, S.F.; Clancy, K.M. Measurement of acid polysaccharides in marine and freshwater samples using alcian blue. *Limnol. Oceanogr. Methods* **2007**, *5*, 73–87. [CrossRef]
38. Villacorte, L.O.; Kennedy, M.D.; Amy, G.L.; Schippers, J.C. The fate of Transparent Exopolymer Particles (TEP) in integrated membrane systems: Removal through pre-treatment processes and deposition on reverse osmosis membranes. *Water Res.* **2009**, *43*, 5039–5052. [CrossRef] [PubMed]
39. Fatibello, A.; Henriques, S.H.S.; Vieira, A.A.; Fatibello-Filho, O. A rapid spectrophotometric method for the determination of transparent exopolymer particles (TEP) in freshwater. *Talanta* **2004**, *62*, 81–85. [CrossRef]
40. Engel, A. The role of transparent exopolymer particles (TEP) in the increase in apparent particle stickiness (alpha) during the decline of a diatom bloom. *J. Plankton Res.* **2000**, *22*, 485–497. [CrossRef]
41. Discard, V.; Bilad, M.R.; Vankelecom, I.F.J. Critical evaluation of the determination methods for transparent exopolymer par-ticles, agents of membrane fouling. *Crit. Rev. Environ. Sci. Technol.* **2015**, *45*, 167–192. [CrossRef]
42. Ohmine, I.; Tanaka, T. Salt effects on the phase transition of ionic gels. *J. Chem. Phys.* **1982**, *77*, 5725–5729. [CrossRef]
43. Benner, R.; Pakulski, J.D.; Mccarthy, M.; Hedges, J.I.; Hatcher, P.G. Bulk Chemical Characteristics of Dissolved Organic Matter in the Ocean. *Science* **1992**, *255*, 1561–1564. [CrossRef]
44. Hansell, D.A.; Carlson, C.A. Deep-ocean gradients in the concentration of dissolved organic carbon. *Nat. Cell Biol.* **1998**, *395*, 263–266. [CrossRef]
45. Hedges, J.I.; Keil, R.G. Sedimentary organic matter preservation: An assessment and speculative synthesis. *Mar. Chem.* **1995**, *49*, 81–115. [CrossRef]
46. Ding, Y.-X.; Hung, C.-C.; Santschi, P.H.; Verdugo, P.; Chin, W.-C. Spontaneous Assembly of Exopolymers from Phytoplankton. *Terr. Atmos. Ocean. Sci.* **2009**, *20*, 741. [CrossRef]
47. Ding, Y.-X.; Chin, W.-C.; Rodriguez, A.; Hung, C.-C.; Santschi, P.H.; Verdugo, P. Amphiphilic exopolymers from Sagittula stellata induce DOM self-assembly and formation of marine microgels. *Mar. Chem.* **2008**, *112*, 11–19. [CrossRef]
48. Bar-On, Y.M.; Phillips, R.; Milo, R. The biomass distribution on Earth. *Proc. Natl. Acad. Sci. USA* **2018**, *115*, 6506–6511. [CrossRef] [PubMed]
49. Decho, A.W. Microbial exopolymer excretions in ocean environments: Their roles in food webs and marine processes. *Oceanogr. Mar. Biol. Annu. Rev.* **1990**, *28*, 73–153.
50. Stoderegger, K.E.; Herndl, G.J. Dynamics in bacterial cell surface properties assessed by fluorescent stains and confocal laser scanning microscopy. *Aquat. Microb. Ecol.* **2004**, *36*, 29–40. [CrossRef]
51. Israelachvili, J.N. *Intermolecular and Surface Forces*; Academic Press: Cambridge, MA, USA, 1992; p. 278.
52. Brochard, F.; de Gennes, P.G. Kinetic of polymer dissolution. *Phys. Chem. Hydrodynam.* **1983**, *4*, 312–322.
53. Orellana, M.V.; Matrai, P.A.; Leck, C.; Rauschenberg, C.D.; Lee, A.M.; Coz, E. Marine microgels as a source of cloud conden-sation nuclei in the high Arctic. *Proc. Natl. Acad. Sci. USA* **2011**, *108*, 13612–13617. [CrossRef]
54. Kato, E. Volume-phase transition of N-isopropyl acrylamide gels induced by hydrostatic pressure. *J. Chem. Phys.* **1997**, *106*, 3792–3798. [CrossRef]
55. Verdugo, P.; Orellana, M.V.; Chin, W.-C.; Petersen, T.W.; Eng, G.V.D.; Benner, R.; Hedges, J.I. Marine biopolymer self-assembly: Implications for carbon cycling in the ocean. *Faraday Dis.* **2008**, *139*, 393–398. [CrossRef] [PubMed]
56. Ding, Y.-X.; Chin, W.-C.; Verdugo, P. Development of a fluorescence quenching assay to measure the fraction of organic carbon present in self-assembled gels in seawater. *Mar. Chem.* **2007**, *106*, 456–462. [CrossRef]
57. Charette, M.A.; Smith, W.H.F. The Volume of Earth's Ocean. *Oceanography* **2010**, *23*, 112–114. [CrossRef]
58. Santschi, P.H.; Guo, L.; Walsh, I.D.; Quigley, M.S.; Baskaran, M. Boundary exchange and scavenging of radionucleotide in continental margin waters of the Middle Atlantic Bight. Implications for organic carbon fluxes. *Cont. Shelf Res.* **1999**, *19*, 609–636. [CrossRef]
59. Santschi, P.H.; Guo, L.; Baskaran, M.; Trumbore, S.; Southon, J.; Bianchi, T.; Honeyman, B.; Cifuentes, L. Isotopic evidence for the contemporary origin of high-molecular weight organic matter in oceanic environments. *Geochim. Cosmochim. Acta* **1995**, *59*, 625–631. [CrossRef]
60. Simon, M.; Grossart, H.; Schweitzer, B.; Ploug, H. Microbial ecology of organic aggregates in aquatic ecosystems. *Aquat. Microb. Ecol.* **2002**, *28*, 175–211. [CrossRef]
61. Mari, X.; Kiorbe, T. Abundance, size distribution and bacterial colonization of transparent exopolymeric particles (TEP) during spring in the Kattegat. *J. Plankton Res.* **1996**, *18*, 969–986. [CrossRef]

62. Whitman, W.B.; Coleman, D.; Wiebe, W.J. Prokaryotes: The unseen majority. *Proc. Natl. Acad. Sci. USA* **1998**, *95*, 6578–6583. [CrossRef]
63. Benner, R.; Amon, R.M. The Size-Reactivity Continuum of Major Bioelements in the Ocean. *Annu. Rev. Mar. Sci.* **2015**, *7*, 185–205. [CrossRef]
64. Rubinstein, M.; Dobrin, A.V. Associations leading to formation of reversible networks and gels. *Curr. Opin. Colloid Interface Sci.* **1999**, *4*, 83–87. [CrossRef]
65. Sun, L.; Chin, W.C.; Chiu, M.-H.; Xu, C.; Lin, P.; Schwehr, K.A.; Quigg, A.; Santschi, P.H. Sunlight induced aggregation of 573 protein-containing dissolved organic matter in the ocean. *Sci. Total Environ.* **2019**, *654*, 872–877. [CrossRef] [PubMed]
66. Santschi, P.; Chin, W.-C.; Quigg, A.; Xu, C.; Kamalanathan, M.; Lin, P.; Shiu, R.-F. Marine Gel Interactions with Hydrophilic and Hydrophobic Pollutants. *Gels* **2021**, *7*, 83. [CrossRef] [PubMed]
67. Osada, Y.; Khokhlov, A. *Polymer Gels and Networks*; Chapman and Hall/CRC: Boca Raton, FL, USA, 2001.

gels

Review

From Nano-Gels to Marine Snow: A Synthesis of Gel Formation Processes and Modeling Efforts Involved with Particle Flux in the Ocean

Antonietta Quigg [1,*], Peter H. Santschi [2], Adrian Burd [3], Wei-Chun Chin [4], Manoj Kamalanathan [1], Chen Xu [2] and Kai Ziervogel [5]

1 Department of Marine Biology, Texas A&M University at Galveston, Galveston, TX 77553, USA; manojka@tamug.edu
2 Department of Marine and Coastal Environmental Science, Texas A&M University at Galveston, Galveston, TX 77553, USA; santschi@tamug.edu (P.H.S.); xuc@tamug.edu (C.X.)
3 Department of Marine Science, University of Georgia, Athens, GA 30602, USA; adrianb@uga.edu
4 Department of Bioengineering, University of California, Merced, CA 95343, USA; wchin2@ucmerced.edu
5 Institute for the Study of Earth, Oceans and Space, University of New Hampshire, Durham, NH 03824, USA; Kai.Ziervogel@unh.edu
* Correspondence: quigga@tamug.edu

Abstract: Marine gels (nano-, micro-, macro-) and marine snow play important roles in regulating global and basin-scale ocean biogeochemical cycling. Exopolymeric substances (EPS) including transparent exopolymer particles (TEP) that form from nano-gel precursors are abundant materials in the ocean, accounting for an estimated 700 Gt of carbon in seawater. This supports local microbial communities that play a critical role in the cycling of carbon and other macro- and micro-elements in the ocean. Recent studies have furthered our understanding of the formation and properties of these materials, but the relationship between the microbial polymers released into the ocean and marine snow remains unclear. Recent studies suggest developing a (relatively) simple model that is tractable and related to the available data will enable us to step forward into new research by following marine snow formation under different conditions. In this review, we synthesize the chemical and physical processes. We emphasize where these connections may lead to a predictive, mechanistic understanding of the role of gels in marine snow formation and the biogeochemical functioning of the ocean.

Keywords: DOM; marine microgels; marine snow; polymer networks theory; biopolymer self-assembly; primary production; phytoplankton secretion; microbial loop; mathematical modeling

Citation: Quigg, A.; Santschi, P.H.; Burd, A.; Chin, W.-C.; Kamalanathan, M.; Xu, C.; Ziervogel, K. From Nano-Gels to Marine Snow: A Synthesis of Gel Formation Processes and Modeling Efforts Involved with Particle Flux in the Ocean. *Gels* **2021**, 7, 114. https://doi.org/10.3390/gels7030114

Academic Editor: Bjørn Torger Stokke

Received: 11 July 2021
Accepted: 6 August 2021
Published: 9 August 2021

1. Introduction

Global biogeochemical cycling of carbon, nitrogen, and other macro- and micro-elements occurs throughout the water column of the oceans. A fraction of the photosynthetically produced carbon in the sunlit photic zone is modified by biotic processes viz the microbial loop and the biological pump [1–7]. Up to 50% of the organic carbon produced by phytoplankton is thought to be taken up by bacteria, which are subsequently grazed by nanoplanktonic heterotrophic flagellates that drive the flux of material and energy into the food chain [3,6,8]. Bacteria, which solubilize particles and acquire dissolved organic carbon (DOC) and inorganic nutrients, are then grazed upon by protozoa, and are subsequently preyed on by mucus net-makers and small zooplankton, the latter of which function as conduits to higher trophic levels. In this way, the passively settling particles below the photic zone, known as marine snow (Figure 1), are regarded as a primary source of substrate that supports heterotrophic food webs [9,10]. The vertical flux of carbon and nutrients relies on sinking particles [11]. The flux of particulate organic carbon (POC) through sinking marine snow from surface waters declines exponentially due to consumption, with only 1% of the sinking organic material reaching the seafloor [12].

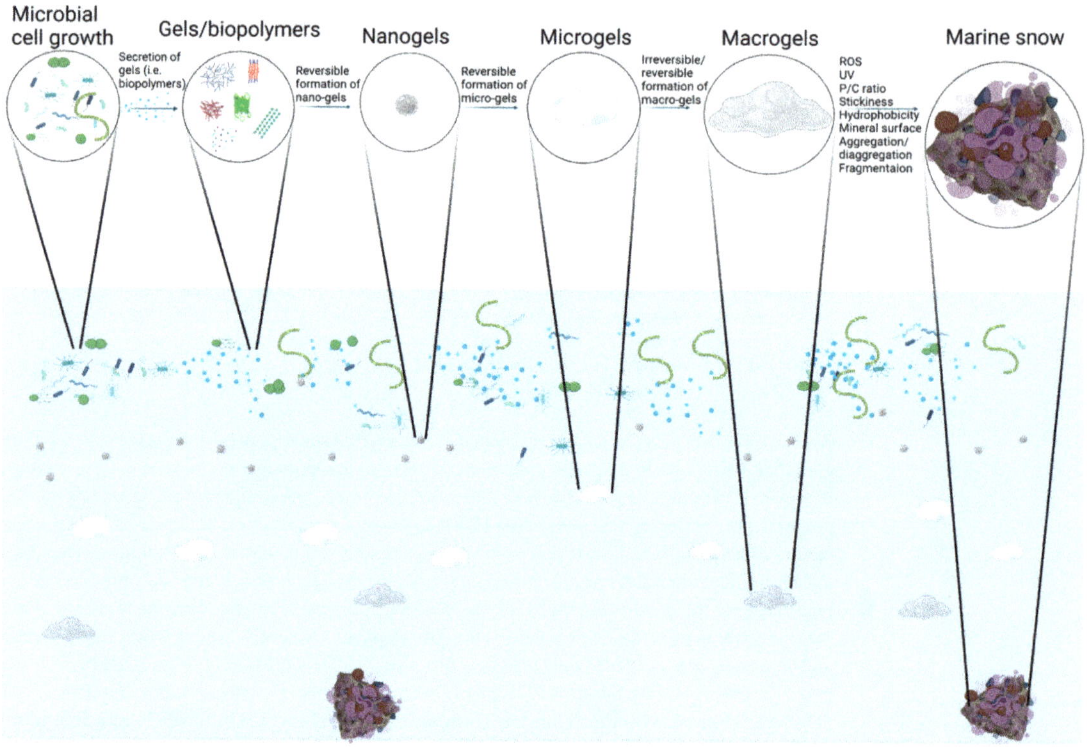

Figure 1. A conceptual model of marine snow formation that requires the following steps: (1) microbial cell growth; (2) secretion of gels (i.e., biopolymers) aka polymeric substances from microbial cells; (3) formation of exopolymeric substances (EPS) which have a variety of forms including TEP and CSP; (4) reversible formation of nano-gels; (5) reversible formation of micro-gels; (5) reversible or irreversible formation of macro-gels; (6) apparent stickiness of particle population dependent on their protein content, i.e., their protein-to-carbohydrate (P/C) ratio; (7) irreversible chemical crosslinking of proteins in gels to form marine snow through hydrophobic or reactive oxygen species (ROS) mediated chemical crosslinking; (8) UV oxidation; (9) interactions of mineral surfaces with gels or marine snow; and (10) aggregation-disaggregation/fragmentation rates. Nanogels (100–150 nm) < microgels (~5 μm) < macrogels (100 μm) < marine snow (>500 μm to 10s of cm) occur on a size continuum.

There is a complicated relationship between DOC and POC, with studies showing a dynamic equilibrium between free and assembled DOC occurring over the whole water column that produces micron-scale gel patchiness that may help to explain carbon turnover, particularly in the dark ocean [13,14]. In addition, Arrieta et al. [15] found that DOC is as readily consumed by bacteria in the surface as in the deep ocean; with rates constrained only by the availability of these materials. A generic relationship between DOC and organic biopolymers forming exopolymeric substances (EPS) [16,17] or transparent exopolymeric particles (TEP) [5,18–20] and larger marine snow composites [21–23] has been suggested [24] but not yet objectively verified. The goal of this review is to synthesize historical and recent literature to examine the relationship (if one exists) between biopolymers released by microbes and marine snow (Figure 1). This is one of the major gaps in our understanding of the mechanisms that lead to marine snow formation and our ability to accurately model particle processes and fluxes in the ocean. Recent studies suggest developing a (relatively) simple model that is tractable and related to the available data; this will enable us to step forward into new research by following marine snow formation under different conditions. This is critical given the variety of anthropogenic factors that are modifying biogeochemical cycles in the marine environment, specifically those whose fate and transport is intrinsically linked to marine snow formation. This includes but is not limited to engineered nanoparticles (e.g., [25,26]), oil spills and dispersants (e.g., [27–30]), and nano- and micro-plastics (e.g., [31–34]). In the recent literature, we often *now* see reference to marine oil snow (MOS; [29,35]) and marine plastic snow (MPS; [36,37]) reflecting the increased awareness and studies in this important arena. This review is not intended to be comprehensive but rather a synthesis of studies across a variety of fields. The reader is therefore referred to the many reviews on gels and their role in the ocean's carbon cycle if that is their specific interest (e.g., [4,5,7,13,14,21,23,30,38–43]).

2. Colloidal Nanogels (or Macromolecules) and Microgels

Riley's [44] early observations of particle formation in seawater pointed to the idea of a reversible exchange between dissolved organic matter (DOM) and particulate organic matter (POM). It is now known that the oceans hold approximately 700 Gt of reduced carbon in a variety of forms (Table 1), with approximately 660 Gt C in the form of DOC [45]. A substantial amount of this material is in the form of microscopic gels that are rich in nutrients and readily available to bacterial colonization [4,14,46–49]. DOM itself remains in reversible assembly/dispersion equilibrium with free biopolymers, forming porous self-assembled microgels. These materials originate in the organic material produced by phytoplankton and bacteria that form 3D polymer-hydrogel networks. Operationally DOM is defined as material that passes through a 0.7 or 0.4 μm pore size filter, and thus includes colloidal particles and macromolecules in the filter-passing fraction. Given the continuum of sizes, ultrafilter-passing molecules (aka the truly dissolved fraction if a 1 kDa membrane is used) are retained by the membrane (if the concentration factor is low) and thus taken as the colloidal fraction. This low molecular weight DOM will have enhanced permeation behavior with increasing concentration factor [50].

Table 1. Major carbon pools in biogeochemical cycles along with their operational definitions.

Acronym	Carbon Pool	Operational Definition
DOC	dissolved organic carbon	Fraction of organic carbon that passes through a filter (either 0.7 μm GF/F or 0.4 μm polycarbonate filter); It contains polymers (e.g., carbohydrates, proteins) that spontaneously self-assemble; these free biopolymers form nanogels.
DOM	dissolved organic matter	Material that passes through a filter (either 0.7 μm GF/F or 0.4 μm polycarbonate filter); includes colloidal particles and macromolecules.
POC	particulate organic carbon	Fraction of carbon retained by the filter.
POM	particulate organic matter	Fraction of organic matter retained by the filter.
EPS	extracellular polymeric substances or exopolymeric substances	Protein and polysaccharide rich materials with smaller amounts of nucleic acids and lipids.
TEP	transparent exopolymeric particles	Alcian-blue stainable transparent particles that are formed from acid polysaccharides.
CSP	Commassie stained particles	Protein-rich Commassie stainable particles.
	marine snow	Composite particles (algae, bacteria, feces) in a matrix of EPS which is visible to the naked eye.

Gel formation and stability depends in part on the physical and chemical properties of their constituent polymers. DOM contains polysaccharides (polyanionic) and proteins (polyelectrolytes) that undergo random motions bringing them into contact with each other, thus resulting in the formation of tangled networks called nanogels, 100–150 μm in size [16,17,45]. Nanogels can be stabilized by Ca^{2+} ions, hydrophobic interactions, or crosslinking by chemical bonds. The rates of polymer collisions, and hence of gel formation depends on the concentration of the polymers and their length, such that gel formation rate depends approximately on the square of the polymer length [51–54]. Gels also disperse through the process of reptation [55]. This occurs when polymers within the gel axially diffuse out of the gel network, thereby becoming free polymers again. This process occurs on timescales that vary with the length of the polymer so that gels formed by long polymers are more stable [56]. Consequently, factors such as UV radiation that can affect polymer length also impact gel formation and stability and the dynamic equilibrium between polymers and gels. The nonlinear behavior of gel formation is described and predicted by polymer gel physics [41,46,49,56–58].

Once nanogels have formed in the oceans, they can interact with each other to form larger, supramolecular networks or microgels (Figure 1) which are ~4–5 μm in size [4,46,58,59] or 3D polymer hydrogel networks [16,60,61] known as physical gels. The latter are stabilized by hydrophobic or ionic bonds as gels are mostly water and so they can interpenetrate [14,49]. This process also leads to a dynamic equilibrium: removal of these "products", i.e., equilibrium goes right for the larger gels (Figure 1), resulting in the formation of new microgels from nanogels, which in turn will lead to the formation of new nanogels as long as there is sufficient concentration of polymers to support their formation. This process also leads to a dynamic equilibrium such that removal of these larger gels will result in the formation of more microgels from nanogels, which in turn will lead to the formation of more nanogels so long as there is sufficient concentration of polymers to support their formation.

Microgel assembly follows a characteristic second-order kinetics with a thermodynamic yield at equilibrium of approximately 10% of the oceans DOC stock, that is, a 10^4 increase of local concentrations of organic material or an estimated 70 Gt of carbon in seawater [14]. Temperature, pressure, and pH, which can vary widely in the ocean, affect DOM self-assembly [14], but salinity does not. Orellana et al. [60] found the polymers in the Artic dissolved organic pool assemble faster and with higher microgel yields than at other latitudes. An important property of gels is that they can undergo phase transitions stimulated by changes in environmental parameters [61]. The presence of molecules such as dimethylsulfoniopropionate and dimethyl-sulfide were found to be critical in the Artic [60]. Changes in these parameters lead to volume phase-transitions (e.g., swelling or dehydration and condensation of the polymer network), collapsing the gel into a denser polymer network. This process can lead to small molecules or even proteins being trapped within the collapsed gel and potentially being transported to depth [62]. The reversible phase transitions shown by these marine gels, as a function of pH, dimethyl-sulfide and dimethylsulfoniopropionate concentrations, can reduce the gel size to <1 μm in diameter [45,60].

Short polymers form only nanogels that remain in continuous assembly/dispersion equilibrium. Dynamic light scattering and other techniques for measuring microgels (up to ~5 μm) have to utilize prefiltered (0.5 μm size) seawater as larger particles interfere. Nevertheless, direct evidence of the critical importance of polymer size on assembly is illustrated by the observation that UV cracking of DOM polymers results in shorter chains, longer assembly times, and smaller-size gels [59]. However, recent studies suggest that if natural sunlight is used, aggregation of EPS by reactive oxygen species (ROS) mediated chemical crosslinking of the protein fraction and microgel formation occurs in parallel to fragmentation and degradation [63–66].

The mix of biopolymers in these gels collectively referred to as EPS includes predominantly polysaccharides (neutral carbohydrates, amino sugars) and proteins [67,68], with nucleic acids and lipids present but at significantly lower concentrations. The distinctive sugar and amino acid compositions of the colloidal fraction are relatively uniform throughout the ocean such that these chemical signatures are used to test for selective assembly of biomacromolecules into gels [57]. These biopolymers are either released by phytoplankton primary production, bacterial activity, or are the end products of the degraded detritus of marine biota [16,39,59]. Acidic polysaccharides such as uronic acids contain carboxyl groups that provide binding sites for divalent (e.g., Ca^{2+}, Mg^{2+}) or trivalent (e.g., Fe^{3+}) ions providing bidentate inner-sphere coordination sites that can cause supra-macromolecular aggregation and Ca^{2+} bridging for structural stability [4]. Proteins, as another major EPS component, are amphiphilic and mediate the stability and aggregation of the 3-D networks of biopolymers, through both hydrophobic and electrostatic interactions [68–70], as well as light-induced cross-linking [63–66]. EPS are often subcategorized into truly dissolved (<1 nm) or colloidal (1–1000 nm) fractions [8,43,58,71–75].

EPS are thought to drive marine particle formation (Figure 1), including marine macrogels and marine snow [9,20,30,74]. EPS may be hydrophobic or hydrophilic [58,75]. Very small amounts of amphiphilic EPS can greatly accelerate micro-gel formation [58]; these in turn have been shown to be biomes for accelerated microbial activity [4,41,76–79]. Stoderegger and Herndl [80] first introduced the idea that hydrophobic interactions might play an important role in coagulation of marine particles, suggesting that EPS hydrophobic properties might be responsible for POC and bacterial aggregation. Studies monitoring DOM assembly showed that EPS from the marine bacteria *Sagittula stellata* induce microgel formation [58,75] following typical hydrophobic polymer-bonding kinetics. The assembly of DOM polymers and hydrophobic EPS follows a first-order kinetics and requires a much lower concentration that are different from DOM polymers or amphiphilic EPS alone. In addition, *Sagittula* EPS-induced DOM assembly and microgel formation exhibited typical temperature-enhanced cooperativity found in hydrophobic interaction-driven processes with typical high cooperativity [75]. Enhancement of hydrophobic interactions with temperature results from temperature-induced conformational changes of amphiphilic polymers that produce increased hydrophobic contact area [69,81] and a higher probability of inter-chain bonding.

An important additional characteristic of EPS-induced DOM assembly is the critical assembly concentration that remains proportional over a broad range of surfactant concentrations and is independent of both polymer charge and the presence of counter ions [81]. In fact, DOM assembly induced by more hydrophobic EPS (higher protein content) can remain virtually unchanged in Ca^{2+}-free seawater [58]. The fundamental mechanisms of hydrophobic bonding within EPS, however, remain obscure [30]. What is clear is that the hydrophobic effect is caused by the interaction (aggregation or clustering) of hydrophobic moieties or molecules (exposed after unfolding) when they are surrounded by hydrophilic environment.

Release of EPS may allow bacteria (and phytoplankton) to wrap themselves in a DOM network of virtually locked-up nutrients. DOM becomes a target easily cleaved by bacterial exoenzymes to yield low-molecular-weight oligomers that can be readily imported and metabolized by the bacteria [79,82,83]. This inference agrees with observations that bacteria are found on hot spots [1,8,78,84,85], lodged in and around the EPS-induced DOM networks or colonizing DOM self-assembled networks [48]. EPS release may therefore allow bacteria to concentrate substrate that is otherwise inaccessible at the characteristic low DOM concentration found in seawater (see also [15]). Recent reports that EPS from *Synechococcus*, *Emiliania huxleyi*, and *Skeletonema costatum* can self-assemble in Ca^{2+}-free artificial seawater indicate that phytoplankton EPS might also be responsible for the production of the vast majority of microgels [75]. Considering the rich and ubiquitous presence of phytoplankton in the ocean and high-yield secretory activity [36,86–88], these results suggest that microalgae might not only be a major source of reduced organic carbon but also may release amphiphiles that can induce DOM assembly.

3. TEP and CSP

Marine gels are part of a colloidal continuum that, at larger sizes, may operationally be defined as TEP or Commasie Stained Particles (CSP) because of their gel-like nature. The work by Alice Alldredge and colleagues first described the link between polysaccharide-containing macrogels (TEP; Figure 1) and the flocculation of diatoms resulting in the formation of marine snow [21,89]. The later studies showed that TEP form discrete sheets, films, or strings ranging from three to 100 μm. The promotion of coagulation of TEP provides the matrix for marine snow [4,39,40,89], particularly when it interacts with calcium or silica based biominerals [90,91]. Today we know that TEP are found throughout the water column from the sunlit surface layers to the dark ocean [92].

Busch et al. [93] and Engel et al. [94] showed detailed distributions of polysaccharides and proteins in marine gel particles that were stained as TEP and CSP, respectively. Busch et al. [93] reported that at all stations, their results showed strong positive correlations over depth between gel particle number and total gel particle area for both gel particle types (TEP and CSP), with gel particle diameters of several hundred μm, and bacterial colonization of 1–2 $\times$ 10^5 cells/mm^2. Average proportions of bacteria attached to gel particles ranged from 1–4%, with most of the bacteria free-living.

Engel et al. [94] state in their paper, "extracting three-dimensional, fractal gel particles onto membrane filters, staining, and subsequent measurement of their two dimensional size introduces inevitable inaccuracies. Likewise, estimating their main chemical components by compound-specific stains that may target one but not all components is prone to error". Regardless of the potential inaccuracies or biases, these gel particles accounted for about 0.1 to 10% of DOC, depending on the depth and ocean. Thus, it seems to be clear that these large gel particles are important, especially in the surface ocean.

It should be noted that TEP and CSP are measured using chemical stains (Alcian blue or Commasie blue respectively at low pH) that potentially alter the physical and/or chemical structures of marine organic matter. These induced alterations discourage further investigation of gel physicochemical characteristics for microgels as described in Chin et al. [46]. Measurements of TEP, EPS, and microgels were recently evaluated in Xu et al., [67]. While there was reasonable agreement or significant relationships between each of the three operational methods, i.e., TEP, EPS, and gels; they are not completely equivalent. Although these terms are often used interchangeably, they are mostly operational definitions given their respective quantification methods [30]. Measurements from one method cannot be directly applied to or substitute for the others. This clarification is critical as it has not been shown that these rigidly fixed and stained particles have the emergent physicochemical properties of gels as defined by Chin et al. [46] using polymer physics approaches. Nor has it been clarified if any staining method might even induce aggregation by crosslinking some macromolecules. An awareness of these differences is fundamentally critical because it affects future modeling efforts (see Section 5 below) if we expect them to be mechanistic and predictive.

4. Marine Snow

How marine snow-sized (particles >500 μm in length) macrogels reach their large final equilibrium size has not been rigorously established (Figure 1). It is likely that they undergo multiple annealing steps whereby interpenetration by elongated undegraded polymers allows these gels to reach a larger, more stable equilibrium size [20,30]. Conversely, assembly of shorter chains from the DOM fraction yields a correspondingly large pool of gel that can self-assemble throughout the whole water column [49].

The gel component of detrital marine snow particles can determine the properties of these particles. For example, Alldredge and Gottschalk [95] determined that marine snow sinking rates were non-Stokesian. Sinking rates were not related to excess density, but to diameter as a power function, with an exponent of about 0.26. In Alldredge et al. [96], they established that marine snow particle aggregate size was related to TEP content. This is likely a result of the positive buoyancy of TEP reducing the excess density of the particle, with larger aggregates containing more TEP and thereby reducing the sinking speed of the aggregates [5].

Changes in surface tension due to emulsifier or EPS addition to seawater also affect or control aggregation processes. Schwehr et al. [81] using model EPS constituents, such as protein (BSA, bovine serum albumin) and uronic acids (glucuronic acid with carboxyl moieties; carrageenan with sulfate groups), showed that increasing the protein to carbohydrate (P/C) ratio of EPS lowers the surface tension, which in-turn resulted in aggregation through Ca^{2+} bridging. This implies that gel growth and marine snow formation occur through a combination of Ca^{2+} bridging of acid polysaccharides, and ROS-mediated chemical crosslinking of proteins, in parallel to enzymatic cleavage, oxidation, and degradation pathways [66]. Schwehr et al. [81] also showed that surface tension overall was a function of the P/C ratio of the EPS in the water. The implication is that the surface tension reduction is related to the gel formation mechanisms.

The ratio of EPS has been determined to be an important parameter in determining the "stickiness" or attachment propensity or aggregation potential [68,97], thereby regulating the overall DOM-particulate organic matter (POM) continuum [4,98]. Among the aggregation mechanisms that need to be considered, there is the newly discovered coupled process of ROS-mediated chemical crosslinking of proteins under sunlight [63–66]. These authors demonstrated that sunlight can directly induce aggregation of different kinds of EPS, with greater effect on EPS containing more protein, [65], as the aggregation of EPS is actually caused by aggregation of proteins, in parallel to some cleavage of high molecular weight compounds into smaller, less stable fragments. The findings revealed that the UVB light is of higher energy that can cleave DOM polymers [46,59] and TEP particles [69], and thus reduces their spontaneous assembly. In contrast, the UVA and visible light have lower energy to cleave polymers, but are capable to induce chemical crosslinking of individual macromolecules and marine snow aggregate formation. The increase in particle size of EPS from seven different microbial species was found to be positively correlated to the P/C ratio of EPS [88]. On the other hand, no marine snow aggregates were observed for a non-protein containing EPS from a phytoplankton species [36,88]. It was also shown that hydroxyl radical and peroxide played critical roles in this photo-oxidation process, and ionic strength (controlled by salt concentration) and Ca^{2+}-bridging assisted the aggregation process that leads to marine snow formation. The formation of higher molecular weight products compared to the native proteins, with simultaneously increased carbonyl content, was demonstrated by gel electrophoresis. The model proteins ultimately became more resistant to proteolysis [66]. The addition of ROS (i.e., H_2O_2 and •OH) only accelerated the observed transformations under simulated sunlight. Sun et al. [66] thus demonstrated that photo-oxidation can transform labile proteinaceous materials into refractory matter, providing a novel mechanism for the preservation of high molecular weight dissolved organic nitrogen in the ocean. These observations provide new insights into polymer assembly, marine snow formation, and the fate/transport of organic carbon and nitrogen in the ocean. The relatively elevated P/C ratio of EPS induced by environmental stresses was found to associate with gel formation [36,88]. The effect of environmental stresses is similar on both phytoplankton and bacteria [88]. The stickiness increase of EPS association with environmental stresses might provide additional insights for the recurrent massive mucus aggregates (sea snot) incidents in Adriatic and Mediterranean Sea [99,100].

Even though the bulk P/C ratios that are used as a proxy for the stickiness of biopolymers that are mainly comprised of the two major components (proteins and carbohydrates) [37,88,97], one still needs to take the individual monosaccharides or amino acid composition (i.e., individual species and their relative abundances) into account for more accurate analysis of the relative hydrophobicity/hydrophilicity of the biopolymers. For example, at neutral pH (6.8), acidic carbohydrates (i.e., glucuronic acid) are negatively charged while amino sugars like glucosamine are positively charged, which results in opposite electrostatic interactions. Another example is the microbial degradation of agal-derived EPS resulted in an increase of the deoxy sugars, fucose, and rhamnose and thus also a possible increase in the hydrophobic features of the EPS due to the carbon six methyl group. Amino acids also have a good grouping according to their hydrophobicity/hydrophilicity scaling [101,102].

A synthesis of the literature, microscopic observations of natural colloids, experimental results obtained with model systems, and numerical simulations, led Buffle et al. [98] to conclude that the formation of aggregates such as marine snow in aquatic systems can be understood by mainly considering the roles of three types of colloids: (i) compact inorganic colloids; (ii) large, rigid biopolymers such as polysaccharides; and (iii) either the soil-derived fulvic compounds or their equivalent in pelagic waters, aquagenic refractory organic matter. In most natural aquatic systems, the small (few nanometers) fulvic compounds will stabilize the inorganic colloids whereas the rigid biopolymers (0.1–1 μm) will destabilize them, i.e., lead to floc (marine snow) formation. The concentration of stable or unstable (i.e., aggregated) colloids in a particular aquatic system will depend on the relative proportions of these three components. Santschi et al. [17] showed experimental evidence that the number of marine snow particles >1.5 mm (imaged by a well-calibrated camera system) in the Middle Atlantic Bight, ranged between 5 and 40 aggregates/L. The abundance of marine snow aggregates (but not the suspended particle concentration) were well correlated with the deficiency of a short-lived particle-reactive radionuclide, Th-234, with respect to its production rate by its long-lived radioactive parent, U-238, down to 2500 m water depth, indicating the strong scavenging ability of these marine snow particle aggregates.

5. Modeling Efforts

Basin-wide and global biogeochemical models of oceanic carbon export often explicitly include DOM, but do not unequivocally include gels or TEP, as the latter two terms are more varied in the field or laboratory collection and detection methods (see above, 67 and references therein) though some take steps to incorporate simplified representations. One of the first numerical models on aggregate formation in aquatic systems was published by Jackson [103] where he combined kinetic coagulation theory with simple algal growth kinetics to describe the dynamics of an algal bloom. Results of his model show that coagulation, and subsequent sinking of aggregates, dominates the dynamics when algal concentrations are above a certain critical value [103,104]. This critical algal concentration varies inversely with the fluid shear, algal size, and a parameter called the "stickiness". The stickiness, or coagulation efficiency, represents the probability that two particles will adhere once they have collided, and it is through this parameter that the effects of gels can potentially be incorporated. However, it is important to note that at present there has been little effort into making the stickiness more than a fudge factor, whereas in reality it is likely related to the formation and properties of EPS, gels, and TEP.

Jackson [105] used his previous coagulation model to examine the results of a mesocosm experiment suggesting that aggregating TEP particles with algal particles changes the overall stickiness of the aggregates. However, it was unclear if changes in the TEP-algal particle interactions resulted from the inherent stickiness of TEP or from the increased particle concentration that resulted from explicitly including the TEP particles in the model. Mari and Burd [18] adapted Jackson's model to explicitly model the coagulation of algal cells with TEP using measured TEP size distributions, but even there, stickiness parameters for the interaction between TEP and algal particles were held fixed and were not based on any underlying physico-chemical processes. Oguz [106] took a different approach by explicitly modeling the formation of TEP from DOM secreted by phytoplankton and bacteria, but prescribing aggregation rates rather than determining them using coagulation theory. More complicated models of marine snow formation have been used that include simplified models of TEP formation and particle aggregation [107,108]. However, even in these models, the stickiness of particles is prescribed and not based on any fundamental quantity such as the P/C ratio which may provide a way of improving aggregation models.

More recently, the PISCES biogeochemical model [109] includes a single, semi-labile DOC pool formed from bacterial, phytoplankton secretion, and zooplankton excretion. The DOC can aggregate with itself to form POC, and it can aggregate with POC. Aggregation in this model assumes that particles obey a steady-state power-law size distribution, which makes the models computationally efficient, but does not fully capture the reversible equilibrium dynamics between DOC, gels, and POC. On the other hand, Maerz et al. [110] modeled particle fluxes in the global ocean using the latest algorithms, called "novel Microstructure, Multiscale, Mechanistic, Marine Aggregates in the Global 5 Ocean (M4AGO) sinking scheme". In their model, they related adhesion properties, i.e., particle stickiness, to the fractal nature of the aggregates, assuming that the stronger the surface adhesive forces are, the higher the stickiness of particles. Accompanying this, the intrusion of particles and particle clusters into each other would also diminish resulting in a looser structure with smaller fractal dimension.

6. Marine Gels and the Ocean's Carbon Cycle

Massive sedimentation events of marine snow have often been observed at the decline of a phytoplankton bloom when the EPS production increases in response to nutrient stress [39,111,112]. Sinking marine snow ultimately removes CO_2 from the atmosphere, thus balancing the atmospheric carbon levels on geologic timescales [12,113]. Quantifying vertical fluxes of POM in the ocean is however, complicated by seasonal and inter-annual variations that determine marine snow formation and sedimentation [114,115]. This in turn alters the quality and quantity of POM, and settling rates, which are a function of the size and density of marine snow [42]. Particulate inorganic matter including clay minerals from terrestrial sources as well as carbonate sheathes in the form of calcite foraminifera shells, coccolith plates, and aragonite pteropod shells and shell fragments, may act as ballast for marine snow increasing its density and thus sinking rates through the water column [11,116]. Additional factors affecting the fate of sinking marine snow include microbial decomposition and grazing as well as physical fragmentation during transit [117,118].

Climate change and related acidification and warming of the surface ocean affect microbial metabolic rates including the release of extracellular organic compounds that form particulate EPS [119–123] and, as a consequence, marine snow sedimentation and carbon sequestration at depth in the future ocean [124]. On the other hand, accelerated microbial oxidation rates of EPS in a warming ocean may counteract carbon export fluxes to the deep sea, preserving more organic matter in surface waters [120,123]. EPS dynamics in a warmer and more acidic ocean have been found to also depend on other environmental factors such as nutrient availability for primary and secondary producers [125,126], complicating future predictions of marine snow sedimentation and thus the efficiency of the biological pump under future climate scenarios.

7. Conclusions

The formation of marine snow from biopolymers secreted by microbes is a dynamic progression that relies on biological, physical, and chemical processes. Biopolymers can associate in a reversible manner to form nanogels, which in-turn can reversibly interact with other supramolecular networks (e.g., geopolymers like humic substances, terrestrial origin and more degraded compared to the freshly produced EPS) [127,128] and DOM to form micro and macrogels. Such dynamic interaction between biopolymers, gels and debris along with the influence from physical and chemical processes such as UV and ROS interactions could eventually leads to the formation of marine snow. Furthermore, factors such as biopolymer interactions with divalent and trivalent ions, as well as their composition (P/C), concentration, and length have been shown to play a major role in the transformation of gels to marine snow. Although, the above mentioned processes and factors to some extent explain the relationship between gels and marine snow, much remains unclear. Further studies on marine snow formation that integrates laboratory and in situ aspects along with incorporation of stickiness factor such as P/C ratio of EPS in modelling efforts can further our understanding of the relationship between gels and marine snow.

Author Contributions: All authors contributed equally to the article and approved the submitted version. All authors have read and agreed to the published version of the manuscript.

Funding: This research received no external funding.

Acknowledgments: We thank Pedro Verdugo (Dept. of Bioengineering, University of Washington, Friday Harbor Marine Laboratories, Washington, WA, USA) for his kind invitation to contribute to this special issue. The authors thank the Gulf of Mexico Research Initiative for their support from 2015–2020 which allowed them to work together and have many fruitful conversations that ultimately supported assembling this manuscript. We also thank Jessica Hillhouse (Texas A&M University at Galveston) for formatting the references for this paper. We thank for thoughtful reviewers for their comments which improved the manuscript.

Conflicts of Interest: The authors declare no conflict of interest.

References

1. Azam, F.; Fenchel, T.; Field, J.G.; Gray, J.S.; Meyer-Reil, L.A.; Thingstad, F. The ecological role of water-column microbes in the sea. *Mar. Ecol. Prog. Ser.* **1983**, *10*, 257–263. [CrossRef]
2. Falkowski, P.G.; Barber, R.T.; Smetacek, V. Biogeochemical controls and feedbacks on ocean primary production. *Science* **1998**, *281*, 200–206. [CrossRef] [PubMed]
3. Ducklow, H.W.; Steinberg, D.K.; Buesseler, K.O. Upper ocean carbon export and the biological pump. *Oceanography* **2001**, *14*, 50–58. [CrossRef]
4. Verdugo, P.; Alldredge, A.L.; Azam, F.; Kirchman, D.L.; Passow, U.; Santschi, P.H. The oceanic gel phase: A bridge in the DOM–POM continuum. *Mar. Chem.* **2004**, *92*, 67–85. [CrossRef]
5. Mari, X.; Passow, U.; Migon, C.; Burd, A.B.; Legendre, L. Transparent exopolymer particles: Effects on carbon cycling in the ocean. *Prog. Oceanogr.* **2017**, *151*, 13–37. [CrossRef]
6. Polimene, L.; Sailley, S.; Clark, D.; Mitra, A.; Allen, J.I. Biological or microbial carbon pump? The role of phytoplankton stoichiometry in ocean carbon sequestration. *J. Plankton Res.* **2017**, *39*, 180–186. [CrossRef]
7. Orellana, M.V.; Leck, C. Chapter 9–Marine microgels. In *Biogeochemistry of Marine Dissolved Organic Matter*, 2nd ed.; Hansell, D.A., Carlson, C.A., Eds.; Academic Press: Boston, MA, USA, 2015; pp. 451–480. [CrossRef]
8. Azam, F.; Smith, D.C.; Steward, G.F.; Hagström, Å. Bacteria-organic matter coupling and its significance for oceanic carbon cycling. *Micro. Ecol.* **1994**, *28*, 167–179. [CrossRef] [PubMed]
9. Decho, A.W. Microbial exopolymer secretions in ocean environments: Their role (s) in food webs and marine processes. *Oceanogr. Mar. Biol. Annu. Rev.* **1990**, *28*, 73–153.
10. Buesseler, K.O.; Lamborg, C.H.; Boyd, P.W.; Lam, P.J.; Trull, T.W.; Bidigare, R.R.; Bishop, J.K.; Casciotti, K.L.; Dehairs, F.; Elskens, M.; et al. Revisiting carbon flux through the ocean's twilight zone. *Science* **2007**, *316*, 567–570. [CrossRef]
11. Armstrong, R.A.; Lee, C.; Hedges, J.I.; Honjo, S.; Wakeham, S.G. A new, mechanistic model for organic carbon fluxes in the ocean based on the quantitative association of POC with ballast minerals. *Deep Sea Res. II Top. Stud. Oceanogr.* **2001**, *49*, 219–236. [CrossRef]
12. Martin, J.H.; Knauer, G.A.; Karl, D.M.; Broenkow, W.W. VERTEX: Carbon cycling in the northeast Pacific. *Deep Sea Res. Part A Oceanogr. Res. Pap.* **1987**, *34*, 267–285. [CrossRef]
13. Verdugo, P.; Santschi, P.H. Polymer dynamics of DOC networks and gel formation in seawater. *Deep Sea Res. Part II Top. Stud. Oceanogr.* **2010**, *57*, 1486–1493. [CrossRef]
14. Verdugo, P. Marine microgels. *Ann. Rev. Mar. Sci.* **2012**, *15*, 4375–4400. [CrossRef]
15. Arrieta, J.M.; Mayol, E.; Hansman, R.L.; Herndl, G.J.; Dittmar, T.; Duarte, C.M. Dilution limits dissolved organic carbon utilization in the deep ocean. *Science* **2015**, *348*, 331–333. [CrossRef] [PubMed]
16. Santschi, P.H.; Balnois, E.; Wilkinson, K.J.; Zhang, J.; Buffle, J.; Guo, L. Fibrillar polysaccharides in marine macromolecular organic matter as imaged by atomic force microscopy and transmission electron microscopy. *Limnol. Oceanogr.* **1998**, *43*, 896–908. [CrossRef]
17. Santschi, P.H.; Guo, L.; Walsh, I.D.; Quigley, M.S.; Baskaran, M. Boundary exchange and scavenging of radionuclides in continental margin waters of the Middle Atlantic Bight: Implications for organic carbon fluxes. *Cont. Shelf Res.* **1999**, *19*, 609–636. [CrossRef]
18. Mari, X.; Burd, A. Seasonal size spectra of transparent exopolymeric particles (TEP) in a coastal sea and comparison with those predicted using coagulation theory. *Mar. Ecol. Prog. Ser.* **1998**, *163*, 63–76. [CrossRef]
19. Zhou, J.; Mopper, K.; Passow, U. The role of surface-active carbohydrates in the formation of transparent exopolymer particles by bubble adsorption of seawater. *Limnol. Oceanogr.* **1998**, *43*, 1860–1871. [CrossRef]

20. Passow, U. Formation of transparent exopolymer particles, TEP, from dissolved precursor material. *Mar. Ecol. Prog, Ser.* **2000**, *192*, 1–11. [CrossRef]
21. Alldredge, A.L.; Silver, M.W. Characteristics, dynamics and significance of marine snow. *Prog. Oceanogr.* **1988**, *20*, 41–82. [CrossRef]
22. Alldredge, A.L.; Jackson, G.A. Aggregation in marine systems. *Deep-Sea Res. Part II Top. Stud. Oceanogr.* **1995**, *42*, 1–7. [CrossRef]
23. Jackson, G.A.; Burd, A.B. Aggregation in the marine environment. *Environ. Sci. Technol.* **1998**, *32*, 2805–2814. [CrossRef]
24. Leppard, G.G.; West, M.M.; Flannigan, D.T.; Carson, J.; Lott, J.N. A classification scheme for marine organic colloids in the Adriatic Sea: Colloid speciation by transmission electron microscopy. *Can. J. Fish. Aq. Sci.* **1997**, *54*, 2334–2349. [CrossRef]
25. Navarro, E.; Baun, A.; Behra, R.; Hartmann, N.B.; Filser, J.; Miao, A.J.; Quigg, A.; Santschi, P.H.; Sigg, L. Environmental behavior and ecotoxicity of engineered nanoparticles to algae, plants, and fungi. *Ecotoxicology* **2008**, *17*, 372–386. [CrossRef] [PubMed]
26. Quigg, A.; Chin, W.C.; Chen, C.S.; Zhang, S.; Jiang, Y.; Miao, A.J.; Schwehr, K.A.; Xu, C.; Santschi, P.H. Direct and indirect toxic effects of engineered nanoparticles on algae: Role of natural organic matter. *ACS Sustain. Chem. Eng.* **2013**, *1*, 686–702. [CrossRef]
27. Chanton, J.; Zhao, T.; Rosenheim, B.E.; Joye, S.; Bosman, S.; Brunner, C.; Yeager, K.M.; Diercks, A.R.; Hollander, D. Using natural abundance radiocarbon to trace the flux of petrocarbon to the seafloor following the Deepwater Horizon oil spill. *Environ. Sci. Technol.* **2015**, *49*, 847–854. [CrossRef]
28. Daly, K.L.; Passow, U.; Chanton, J.; Hollander, D. Assessing the impacts of oil-associated marine snow formation and sedimentation during and after the Deepwater Horizon oil spill. *Anthropocene* **2016**, *13*, 18–33. [CrossRef]
29. Passow, U.; Ziervogel, K. Marine snow sedimented oil released during the Deepwater Horizon spill. *Oceanography* **2016**, *29*, 118–125. [CrossRef]
30. Quigg, A.; Passow, U.; Chin, W.C.; Xu, C.; Doyle, S.; Bretherton, L.; Kamalanathan, M.; Williams, A.K.; Sylvan, J.B.; Finkel, Z.V.; et al. The role of microbial exopolymers in determining the fate of oil and chemical dispersants in the ocean. *Limnol. Oceanogr. Lett.* **2016**, *1*, 3–26. [CrossRef]
31. Michels, J.; Stippkugel, A.; Lenz, M.; Wirtz, K.; Engel, A. Rapid aggregation of biofilm-covered microplastics with marine biogenic particles. *Proc. R. Soc. B* **2018**, *285*, 20181203. [CrossRef]
32. Summers, S.; Henry, T.; Gutierrez, T. Agglomeration of nano-and microplastic particles in seawater by autochthonous and de novo-produced sources of exopolymeric substances. *Mar. Poll. Bull.* **2018**, *130*, 258–267. [CrossRef] [PubMed]
33. Porter, A.; Lyons, B.P.; Galloway, T.S.; Lewis, C. Role of marine snows in microplastic fate and bioavailability. *Environ. Sci. Technol.* **2018**, *52*, 7111–7119. [CrossRef] [PubMed]
34. Kvale, K.F.; Friederike Prowe, A.E.; Oschlies, A. A critical examination of the role of marine snow and zooplankton fecal pellets in removing ocean surface microplastic. *Front. Mar. Sci.* **2020**, *6*, 808. [CrossRef]
35. Passow, U.; Ziervogel, K.; Asper, V.; Diercks, A. Marine snow formation in the aftermath of the Deepwater Horizon oil spill in the Gulf of Mexico. *Environ. Res. Lett.* **2012**, *7*, 035301. [CrossRef]
36. Shiu, R.F.; Vazquez, C.I.; Chiang, C.Y.; Chiu, M.H.; Chen, C.S.; Ni, C.W.; Gong, G.C.; Quigg, A.; Santschi, P.H.; Chin, W.C. Nano- and microplastics trigger secretion of protein-rich extracellular polymeric substances from phytoplankton. *Sci. Total Environ.* **2020**, *748*, 141469. [CrossRef] [PubMed]
37. Santschi, P.H.; Chin, W.-C.; Quigg, A.; Xu, C.; Kamalanathan, M.; Lin, P.; Shiu, R.-F. Marine gel interactions with hydrophilic and hydrophobic pollutants. *Gels* **2021**, *7*, 83. [CrossRef]
38. Žutić, V.; Svetličić, V. *Interfacial Processes in Marine Chemistry*; Springer: Berlin/Heidelberg, Germany, 2000; pp. 149–165.
39. Thornton, D.C. Diatom aggregation in the sea: Mechanisms and ecological implications. *Eur. J. Phycol.* **2002**, *37*, 149–161. [CrossRef]
40. Passow, U. Transparent exopolymer particles (TEP) in aquatic environments. *Prog. Oceanogr.* **2002**, *55*, 287–333. [CrossRef]
41. Verdugo, P. Dynamics of marine biopolymer networks. *Polym. Bull.* **2007**, *58*, 139–143. [CrossRef]
42. Burd, A.B.; Jackson, G.A. Particle aggregation. *Ann. Rev. Mar. Sci.* **2009**, *1*, 65–90. [CrossRef]
43. Decho, A.W.; Gutierrez, T. Microbial extracellular polymeric substances (EPSs) in ocean systems. *Front. Microbiol.* **2017**, *8*, 922. [CrossRef] [PubMed]
44. Riley, G.A. Organic aggregates in seawater and the dynamics of their formation and utilization. *Limnol. Oceanogr.* **1963**, *8*, 372–381. [CrossRef]
45. Hansell, D.A. Recalcitrant dissolved organic carbon fractions. *Ann. Rev. Mar. Sci.* **2013**, *5*, 421–445. [CrossRef] [PubMed]
46. Chin, W.C.; Orellana, M.V.; Verdugo, P. Spontaneous assembly of marine dissolved organic matter into polymer gels. *Nature* **1998**, *391*, 568–572. [CrossRef]
47. Ding, Y.X.; Chin, W.C.; Verdugo, P. Development of a fluorescence quenching assay to measure the fraction of organic carbon present in self-assembled gels in seawater. *Mar. Chem.* **2007**, *106*, 456–462. [CrossRef]
48. Moon, A.; Oviedo, A.; Ng, C.; Tuthill, J.; Dmitrijeva, J. Bacterial colonization of marine gel. In Proceedings of the American Society for Limnology and Oceanography Aquatic Sciences Meeting, Santa Fe, NM, USA, 4–9 February 2007.
49. Verdugo, P.; Orellana, M.V.; Chin, W.C.; Petersen, T.W.; van den Eng, G.; Benner, R.; Hedges, J.I. Marine biopolymer self-assembly: Implications for carbon cycling in the ocean. *Faraday Discuss.* **2008**, *139*, 393–398. [CrossRef] [PubMed]
50. Guo, L.; Wen, L.S.; Tang, D.; Santschi, P.H. Re-examination of cross-flow ultrafiltration for sampling aquatic colloids: Evidence from molecular probes. *Mar. Chem.* **2000**, *69*, 75–90. [CrossRef]

51. Edwards, S.F.; Grant, J.W. The effect of entanglements of diffusion in a polymer melt. *J. Phys. A Math. Nucl. Gen.* **1973**, *6*, 1169. [CrossRef]
52. Edwards, S.F. The dynamics of polymer networks. *J. Phys. A Math. Nucl. Gen.* **1974**, *7*, 318. [CrossRef]
53. Edwards, S.F. The theory of macromolecular networks. *Biorheology* **1986**, *23*, 589–603. [CrossRef]
54. Grosberg, A.Y.; Khokhlov, A.R. *Giant Molecules*; Academic Press: Cambridge, MA, USA, 1997.
55. Doi, M.; Edwards, S.F. *The Theory of Polymer Dynamics*; Oxford University Press: Oxford, UK, 1988.
56. De Gennes, P.G.; Leger, L. Dynamics of entangled polymer chains. *Ann. Rev. Phys. Chem.* **1982**, *33*, 49–61. [CrossRef]
57. Orellana, M.V.; Petersen, T.W.; Diercks, A.H.; Donohoe, S.; Verdugo, P.; van den Engh, G. Marine microgels: Optical and proteomic fingerprints. *Mar. Chem.* **2007**, *105*, 229–239. [CrossRef]
58. Ding, Y.X.; Chin, W.C.; Rodriguez, A.; Hung, C.C.; Santschi, P.H.; Verdugo, P. Amphiphilic exopolymers from *Sagittula stellata* induce DOM self-assembly and formation of marine microgels. *Mar. Chem.* **2008**, *112*, 11–19. [CrossRef]
59. Orellana, M.V.; Verdugo, P. Ultraviolet radiation blocks the organic carbon exchange between the dissolved phase and the gel phase in the ocean. *Limnol. Oceanogr.* **2003**, *48*, 1618–1623. [CrossRef]
60. Orellana, M.V.; Matrai, P.A.; Leck, C.; Rauschenberg, C.D.; Lee, A.M.; Coz, E. Marine microgels as a source of cloud condensation nuclei in the high Arctic. *Proc. Natl. Acad. Sci. USA* **2011**, *108*, 13612–13617. [CrossRef]
61. Tanaka, T.; Fillmore, D.; Sun, S.T.; Nishio, I.; Swislow, G.; Shah, A. Phase transitions in ionic gels. *Phys. Rev. Lett.* **1980**, *45*, 1636. [CrossRef]
62. Orellana, M.N.; Hansell, D.A. Ribulose-1, 5-bisphosphate carboxylase/oxygenase (RuBisCO): A long-lived protein in the deep ocean. *Limnol. Oceanogr.* **2012**, *57*, 826–834. [CrossRef]
63. Sun, L.; Xu, C.; Zhang, S.; Lin, P.; Schwehr, K.A.; Quigg, A.; Chiu, M.H.; Chin, W.C.; Santschi, P.H. Light-induced aggregation of microbial exopolymeric substances. *Chemosphere* **2017**, *181*, 675–681. [CrossRef]
64. Sun, L.; Chiu, M.H.; Xu, C.; Lin, P.; Schwehr, K.A.; Bacosa, H.; Kamalanathan, M.; Quigg, A.; Chin, W.C.; Santschi, P.H. The effects of sunlight on the composition of exopolymeric substances and subsequent aggregate formation during oil spills. *Mar. Chem.* **2018**, *203*, 49–54. [CrossRef]
65. Sun, L.; Chin, W.C.; Chiu, M.H.; Xu, C.; Lin, P.; Schwehr, K.A.; Quigg, A.; Santschi, P.H. Sunlight induced aggregation of dissolved organic matter: Role of proteins in linking organic carbon and nitrogen cycling in seawater. *Sci. Total Environ.* **2019**, *654*, 872–877. [CrossRef]
66. Sun, L.; Xu, C.; Lin, P.; Quigg, A.; Chin, W.C.; Santschi, P.H. Photo-oxidation of proteins facilitates the preservation of high molecular weight dissolved organic nitrogen in the ocean. *Mar. Chem.* **2021**, *229*, 103907. [CrossRef]
67. Xu, C.; Chin, W.C.; Lin, P.; Chen, H.; Chiu, M.H.; Waggoner, D.C.; Xing, W.; Sun, L.; Schwehr, K.A.; Hatcher, P.G.; et al. Comparison of microgels, extracellular polymeric substances (EPS) and transparent exopolymeric particles (TEP) determined in seawater with and without oil. *Mar. Chem.* **2019**, *215*, 103667. [CrossRef]
68. Santschi, P.H.; Xu, C.; Schwehr, K.A.; Lin, P.; Sun, L.; Chin, W.C.; Kamalanathan, M.; Bacosa, H.P.; Quigg, A. Can the protein/carbohydrate (P/C) ratio of exopolymeric substances (EPS) be used as a proxy for their 'stickiness' and aggregation propensity? *Mar. Chem.* **2020**, *218*, 103734. [CrossRef]
69. Ortega-Retuerta, E.; Passow, U.; Duarte, C.M.; Reche, I. Effects of ultraviolet B radiation on (not so) transparent exopolymer particles. *Biogeosciences* **2009**, *6*, 3071–3080. [CrossRef]
70. Song, W.; Zhao, C.; Mu, S.; Pan, X.; Zhang, D.; Al-Misned, F.A.; Mortuza, M.G. Effects of irradiation and pH on fluorescence properties and flocculation of extracellular polymeric substances from the cyanobacterium *Chroococcus minutus*. *Colloids Surf. B Biointerfaces* **2015**, *128*, 115–118. [CrossRef]
71. McCarthy, M.; Hedges, J.; Benner, R. Major biochemical composition of dissolved high molecular weight organic matter in seawater. *Mar. Chem.* **1996**, *55*, 281–297. [CrossRef]
72. Hung, C.C.; Guo, L.; Santschi, P.H.; Alvarado-Quiroz, N.; Haye, J.M. Distributions of carbohydrate species in the Gulf of Mexico. *Mar. Chem.* **2003**, *81*, 119–135. [CrossRef]
73. Hung, C.C.; Guo, L.; Roberts, K.A.; Santschi, P.H. Upper ocean carbon flux determined by the 234Th approach and sediment traps using size-fractionated POC and 234Th data from the Gulf of Mexico. *Geochem. J.* **2004**, *38*, 601–611. [CrossRef]
74. Bhaskar, P.V.; Bhosle, N.B. Microbial extracellular polymeric substances in marine biogeochemical processes. *Curr. Sci.* **2005**, *88*, 45–53.
75. Ding, Y.; Hung, C.C.; Santschi, P.H.; Verdugo, P.; Chin, W.C. Spontaneous assembly of exopolymers from phytoplankton. *Terr. Atmos. Ocean. Sci.* **2009**, *20*, 741–747. [CrossRef]
76. Gutierrez, T.; Teske, A.; Ziervogel, K.; Passow, U.; Quigg, A. Microbial exopolymers: Sources, chemico-physiological properties, and ecosystem effects in the marine environment. *Front. Microbiol.* **2018**, *9*, 1822. [CrossRef]
77. Doyle, S.M.; Whitaker, E.A.; De Pascuale, V.; Wade, T.L.; Knap, A.H.; Santschi, P.H.; Quigg, A.; Sylvan, J.B. Rapid formation of microbe-oil aggregates and changes in community composition in coastal surface water following exposure to oil and the dispersant Corexit. *Front. Microbiol.* **2018**, *9*, 689. [CrossRef]
78. Doyle, S.M.; Lin, G.; Morales-McDevitt, M.; Wade, T.L.; Quigg, A.; Sylvan, J.B. Niche Partitioning between Coastal and Offshore Shelf Waters Results in Differential Expression of Alkane and Polycyclic Aromatic Hydrocarbon Catabolic Pathways. *mSystems* **2020**, *5*, e00668-20. [CrossRef]

79. Kamalanathan, M.; Doyle, S.M.; Xu, C.; Achberger, A.M.; Wade, T.L.; Schwehr, K.; Santschi, P.H.; Sylvan, J.B.; Quigg, A. Exoenzymes as a signature of microbial response to marine environmental conditions. *mSystems* **2020**, *5*, e00290-20. [CrossRef]
80. Stoderegger, K.E.; Herndl, G.J. Dynamics in bacterial cell surface properties assessed by fluorescent stains and confocal laser scanning microscopy. *Aquat. Microb. Ecol.* **2004**, *36*, 29–40. [CrossRef]
81. Xu, C.; Zhang, S.; Chuang, C.Y.; Miller, E.J.; Schwehr, K.A.; Santschi, P.H. Chemical composition and relative hydrophobicity of microbial exopolymeric substances (EPS) isolated by anion exchange chromatography and their actinide-binding affinities. *Mar. Chem.* **2011**, *126*, 27–36. [CrossRef]
82. Schwehr, K.A.; Xu, C.; Chiu, M.H.; Zhang, S.; Sun, L.; Lin, P.; Beaver, M.; Jackson, C.; Agueda, O.; Bergen, C.; et al. Protein: Polysaccharide ratio in exopolymeric substances controlling the surface tension of seawater in the presence or absence of surrogate Macondo oil with and without Corexit. *Mar. Chem.* **2018**, *206*, 84–92. [CrossRef]
83. Diamant, H.; Andelman, D. Onset of self-assembly in polymer-surfactant systems. *EPL (Europhys. Lett.)* **1999**, *48*, 170. [CrossRef]
84. Arnosti, C. Microbial extracellular enzymes and the marine carbon cycle. *Ann. Rev. Mar. Sci.* **2011**, *3*, 401–425. [CrossRef]
85. Kamalanathan, M.; Xu, C.; Schwehr, K.; Bretherton, L.; Beaver, M.; Doyle, S.M.; Genzer, J.; Hillhouse, J.; Sylvan, J.B.; Santschi, P.; et al. Extracellular enzyme activity profile in a chemically enhanced water accommodated fraction of surrogate oil: Toward understanding microbial activities after the Deepwater Horizon oil spill. *Front. Microbiol.* **2018**, *9*, 798. [CrossRef]
86. Chin, W.C.; Orellana, M.V.; Quesada, I.; Verdugo, P. Secretion in unicellular marine phytoplankton: Demonstration of regulated exocytosis in Phaeocystis globosa. *Plant Cell Physiol.* **2004**, *45*, 535–542. [CrossRef]
87. Chiu, M.H.; Khan, Z.A.; Garcia, S.G.; Le, A.D.; Kagiri, A.; Ramos, J.; Tsai, S.M.; Drobenaire, H.W.; Santschi, P.H.; Quigg, A.; et al. Effect of engineered nanoparticles on exopolymeric substances release from marine phytoplankton. *Nanoscale Res. Lett.* **2017**, *12*, 1–7. [CrossRef]
88. Shiu, R.F.; Chiu, M.H.; Vazquez, C.I.; Tsai, Y.Y.; Le, A.; Kagiri, A.; Xu, C.; Kamalanathan, M.; Bacosa, H.P.; Doyle, S.M.; et al. Protein to carbohydrate (P/C) ratio changes in microbial extracellular polymeric substances induced by oil and Corexit. *Mar. Chem.* **2020**, *223*, 103789. [CrossRef]
89. Alldredge, A.L.; Passow, U.; Logan, B.E. The abundance and significance of a class of large, transparent organic particles in the ocean. *Deep Sea Res. Part I Oceanogr. Res. Pap.* **1993**, *40*, 1131–1140. [CrossRef]
90. Brunner, E.; Richthammer, P.; Ehrlich, H.; Paasch, S.; Simon, P.; Ueberlein, S.; van Pée, K.H. Chitin-based organic networks: An integral part of cell wall biosilica in the diatom *Thalassiosira pseudonana*. *Angew. Chem. Int. Ed.* **2009**, *48*, 9724–9727. [CrossRef] [PubMed]
91. Spinde, K.; Kammer, M.; Freyer, K.; Ehrlich, H.; Vournakis, J.N.; Brunner, E. Biomimetic silicification of fibrous chitin from diatoms. *Chem. Mat.* **2011**, *23*, 2973–2978. [CrossRef]
92. Nagata, T.; Yamada, Y.; Fukuda, H. Transparent Exopolymer Particles in Deep Oceans: Synthesis and Future Challenges. *Gels* **2021**, *7*, 75. [CrossRef]
93. Busch, K.; Endres, S.; Iversen, M.H.; Michels, J.; Nöthig, E.M.; Engel, A. Bacterial colonization and vertical distribution of marine gel particles (TEP and CSP) in the Arctic Fram Strait. *Front. Mar. Sci.* **2017**, *4*, 166. [CrossRef]
94. Engel, A.; Endres, S.; Galgani, L.; Schartau, M. Marvelous marine microgels: On the distribution and impact of gel-like particles in the oceanic water-column. *Front. Mar. Sci.* **2020**, *7*, 405. [CrossRef]
95. Alldredge, A.L.; Gotschalk, C. In situ settling behavior of marine snow 1. *Limnol. Oceanogr.* **1988**, *33*, 339–351. [CrossRef]
96. Alldredge, A.L.; Passow, U.; Haddock, H.D. The characteristics and transparent exopolymer particle (TEP) content of marine snow formed from thecate dinoflagellates. *J. Plankton Res.* **1998**, *20*, 393–406. [CrossRef]
97. Chen, C.S.; Shiu, R.F.; Hsieh, Y.Y.; Xu, C.; Vazquez, C.I.; Cui, Y.; Hsu, I.C.; Quigg, A.; Santschi, P.H.; Chin, W.C. Stickiness of extracellular polymeric substances on different surfaces via magnetic tweezers. *Sci. Total Environ.* **2021**, *757*, 143766. [CrossRef]
98. Buffle, J.; Wilkinson, K.J.; Stoll, S.; Filella, M.; Zhang, J. A generalized description of aquatic colloidal interactions: The three-colloidal component approach. *Environ. Sci. Technol.* **1998**, *32*, 2887–2899. [CrossRef]
99. Flander-Putrle, V.; Malej, A. The evolution and phytoplankton composition of mucilaginous aggregates in the northern Adriatic Sea. *Harmful Algae* **2008**, *7*, 752–761. [CrossRef]
100. Danovaro, R.; Fonda, U.S.; Pusceddu, A. Climate change and the potential spreading of marine mucilage and microbial pathogens in the Mediterranean Sea. *PLoS ONE* **2009**, *4*, e7006. [CrossRef]
101. Sereda, T.J.; Mant, C.T.; Sönnichsen, F.D.; Hodges, R.S. Reversed-phase chromatography of synthetic amphipathic α-helical peptides as a model for ligand/receptor interactions Effect of changing hydrophobic environment on the relative hydrophilicity/hydrophobicity of amino acid side-chains. *J. Chromatogr.* **1994**, *676*, 139–153. [CrossRef]
102. Monera, O.D.; Sereda, T.J.; Zhou, N.E.; Kay, C.M.; Hodges, R.S. Relationship of sidechain hydrophobicity and alpha-helical propensity on the stability of the single-stranded amphipathic alpha-helix. *J. Pept. Sci.* **1995**, *1*, 319–329. [CrossRef]
103. Jackson, G.A. A model of the formation of marine algal flocs by physical coagulation processes. *Deep Sea Res. Part A Oceanogr. Res. Pap.* **1990**, *37*, 1197–1211. [CrossRef]
104. Jackson, G.A.; Kiørboe, T. Maximum phytoplankton concentrations in the sea. *Limnol. Oceanogr.* **2008**, *53*, 395–399. [CrossRef]
105. Jackson, G.A. TEP and coagulation during a mesocosm experiment. *Deep Sea Res. Part II Top. Stud. Oceanogr.* **1995**, *42*, 215–222. [CrossRef]
106. Oguz, T. Modeling aggregate dynamics of transparent exopolymer particles (TEP) and their interactions with a pelagic food web. *Mar. Ecol. Prog. Ser.* **2017**, *582*, 15–31. [CrossRef]

107. Jokulsdottir, T.; Archer, D. A stochastic, Lagrangian model of sinking biogenic aggregates in the ocean (SLAMS 1.0): Model formulation, validation and sensitivity. *Geosci. Model Dev.* **2016**, *9*, 1455–1476. [CrossRef]
108. Dissanayake, A.L.; Burd, A.B.; Daly, K.L.; Francis, S.; Passow, U. Numerical modeling of the interactions of oil, marine snow, and riverine sediments in the ocean. *J. Geophys. Res. Ocean.* **2018**, *123*, 5388–5405. [CrossRef]
109. Aumont, O.; Éthé, C.; Tagliabue, A.; Bopp, L.; Gehlen, M. PISCES-v2: An ocean biogeochemical model for carbon and ecosystem studies. *Geosci. Model Dev.* **2015**, *8*, 2465–2513. [CrossRef]
110. Maerz, J.; Six, K.D.; Stemmler, I.; Ahmerkamp, S.; Ilyina, T. Microstructure and composition of marine aggregates as co-determinants for vertical particulate organic carbon transfer in the global ocean. *Biogeoscience* **2020**, *17*, 1765–1803. [CrossRef]
111. Passow, U.; Alldredge, A.L. Aggregation of a diatom bloom in a mesocosm: The role of transparent exopolymer particles (TEP). *Deep Sea Res. Part II Top. Stud. Oceanogr.* **1995**, *42*, 99–109. [CrossRef]
112. Trudnowska, E.; Lacour, L.; Ardyna, M.; Rogge, A.; Irisson, J.O.; Waite, A.M.; Babin, M.; Stemmann, L. Marine snow morphology illuminates the evolution of phytoplankton blooms and determines their subsequent vertical export. *Nat. Commun.* **2021**, *12*, 2816. [CrossRef] [PubMed]
113. Hedges, J.I. Global biogeochemical cycles: Progress and problems. *Mar. Chem.* **1992**, *39*, 67–93. [CrossRef]
114. Lampitt, R.S.; Salter, I.; de Cuevas, B.A.; Hartman, S.; Larkin, K.E.; Pebody, C.A. Long-term variability of downward particle flux in the deep northeast Atlantic: Causes and trends. *Deep Sea Res. Part II Top. Stud. Oceanogr.* **2010**, *57*, 1346–1361. [CrossRef]
115. Smith, K.L.; Ruhl, H.A.; Kahru, M.; Huffard, C.L.; Sherman, A.D. Deep ocean communities impacted by changing climate over 24 y in the abyssal northeast Pacific Ocean. *Proc. Natl. Acad. Sci. USA* **2013**, *110*, 19838–19841. [CrossRef]
116. Passow, U.; De La Rocha, C.L. Accumulation of mineral ballast on organic aggregates. *Glob. Biogeochem. Cycles* **2006**, *20*. [CrossRef]
117. Briggs, N.; Dall'Olmo, G.; Claustre, H. Major role of particle fragmentation in regulating biological sequestration of CO_2 by the oceans. *Science* **2020**, *367*, 791–793. [CrossRef]
118. Collins, J.R.; Edwards, B.R.; Thamatrakoln, K.; Ossolinski, J.E.; DiTullio, G.R.; Bidle, K.D.; Doney, S.C.; Van Mooy, B.A. The multiple fates of sinking particles in the North Atlantic Ocean. *Glob. Biogeochem. Cycles* **2015**, *9*, 1471–1494. [CrossRef]
119. Borchard, C.; Engel, A. Organic matter exudation by *Emiliania huxleyi* under simulated future ocean conditions. *Biogeoscience* **2012**, *9*, 3405–3423. [CrossRef]
120. Endres, S.; Galgani, L.; Riebesell, U.; Schulz, K.G.; Engel, A. Stimulated bacterial growth under elevated p CO_2: Results from an off-shore mesocosm study. *PLoS ONE* **2014**, *9*, e99228. [CrossRef]
121. Engel, A. Direct relationship between CO_2 uptake and transparent exopolymer particles production in natural phytoplankton. *J. Plankton Res.* **2002**, *24*, 49–53. [CrossRef]
122. Liu, J.; Weinbauer, M.G.; Maier, C.; Dai, M.; Gattuso, J.P. Effect of ocean acidification on microbial diversity and on microbe-driven biogeochemistry and ecosystem functioning. *Aquat. Microb. Ecol.* **2010**, *61*, 291–305. [CrossRef]
123. Piontek, J.; Händel, N.; Langer, G.; Wohlers, J.; Riebesell, U.; Engel, A. Effects of rising temperature on the formation and microbial degradation of marine diatom aggregates. *Aquat. Microb. Ecol.* **2009**, *54*, 305–318. [CrossRef]
124. Arrigo, K.R. Marine manipulations. *Nature* **2007**, *450*, 491–492. [CrossRef]
125. Otero, A.; Vincenzini, M. Nostoc (Cyanophyceae) goes nude: Extracellular polysaccharides serve as a sink for reducing power under unbalanced C:N metabolism. *J. Phycol.* **2004**, *40*, 74–81. [CrossRef]
126. Passow, U.; Laws, E.A. Ocean acidification as one of multiple stressors: Growth response of *Thalassiosira weissflogii* (diatom) under temperature and light stress. *Mar. Ecol. Prog. Ser.* **2015**, *541*, 75–90. [CrossRef]
127. Piccolo, A. The supramolecular structure of humic substances. *Soil Sci.* **2001**, *166*, 810–832. [CrossRef]
128. Wells, M.J.; Stretz, H.A. Supramolecular architectures of natural organic matter. *Sci. Total Environ.* **2019**, *671*, 1125–1133. [CrossRef]

gels

Review

Marine Gel Interactions with Hydrophilic and Hydrophobic Pollutants

Peter H. Santschi [1,*], Wei-Chun Chin [2], Antonietta Quigg [3], Chen Xu [1], Manoj Kamalanathan [3], Peng Lin [1] and Ruei-Feng Shiu [4,5]

1 Department of Marine and Coastal Environmental Science, Texas A&M University at Galveston, Galveston, TX 77554, USA; xuc@tamug.edu (C.X.); pengl1104@tamug.edu (P.L.)
2 Department of Bioengineering, University of California, Merced, CA 95343, USA; wchin2@ucmerced.edu
3 Department of Marine Biology, Texas A&M University at Galveston, Galveston, TX 77554, USA; quigga@tamug.edu (A.Q.); manojka@tamug.edu (M.K.)
4 Institute of Marine Environment and Ecology, National Taiwan Ocean University, Keelung 20224, Taiwan; rfshiu@mail.ntou.edu.tw
5 Center of Excellence for the Oceans, National Taiwan Ocean University, Keelung 20224, Taiwan
* Correspondence: santschi@tamug.edu

Abstract: Microgels play critical roles in a variety of processes in the ocean, including element cycling, particle interactions, microbial ecology, food web dynamics, air–sea exchange, and pollutant distribution and transport. Exopolymeric substances (EPS) from various marine microbes are one of the major sources for marine microgels. Due to their amphiphilic nature, many types of pollutants, especially hydrophobic ones, have been found to preferentially associate with marine microgels. The interactions between pollutants and microgels can significantly impact the transport, sedimentation, distribution, and the ultimate fate of these pollutants in the ocean. This review on marine gels focuses on the discussion of the interactions between gel-forming EPS and pollutants, such as oil and other hydrophobic pollutants, nanoparticles, and metal ions.

Keywords: marine gels; aggregates; marine snow; hydrophobic and hydrophilic interactions

Citation: Santschi, P.H.; Chin, W.-C.; Quigg, A.; Xu, C.; Kamalanathan, M.; Lin, P.; Shiu, R.-F. Marine Gel Interactions with Hydrophilic and Hydrophobic Pollutants. *Gels* **2021**, *7*, 83. https://doi.org/10.3390/gels7030083

Academic Editor: Pedro Verdugo

Received: 13 May 2021
Accepted: 27 June 2021
Published: 6 July 2021

1. Introduction

Pollutants in the environment encompass many extraneous substances that, when interacting with natural organic matter (NOM), change their properties as they then become parts of a new, macromolecular, complex. Pollutants are mostly human-made and include hydrophilic metal ions, hydrophobic or amphiphilic low-molecular-weight organic molecules, and nanoparticles, including micro- and nano-plastics. Very often then, these pollutants are 'hitching' a ride with the natural organic molecules, which are composed of terrestrially derived humic and fulvic substances, and microbially secreted EPS. While most of the literature on interactions between metal ions and NOM is devoted to understanding the binding strength, and the extent and kinetics of binding, there is much less known on the nonspecific interactions of metal ions with gel-forming EPS that can modify its gel properties. In this paper, we focus on reviewing the recent literature on interactions between gel-forming EPS and pollutants such as oil and other hydrophobic pollutants, nanoparticles, and metal ions.

EPS are mainly composed of proteins and polysaccharides, as well as smaller amounts of nucleic acids, lipids, and humic substances. EPS make up an important part of NOM in the ocean, in its particulate, colloidal, and macromolecular forms [1]. The plankton–EPS system is a dynamic system, whereby phytoplankton and bacteria form a synergistic relationship in the phycosphere. Phytoplankton secrete photosynthesized carbohydrates and polysaccharides, and associated bacteria degrade some of this material and make available other compounds such as vitamin B12 [2] and hydroxamate siderophores [3,4] to phytoplankton.

The microbial community can regulate the physico-chemical properties of the released EPS in response to changing conditions by secreting [5] polysaccharide-rich EPS (mostly phytoplankton) and protein-rich EPS (mostly bacteria [6]). These biopolymers can interact and bond with each other via ionic forces, van der Waals forces, electrostatic forces, hydrophobic interactions, hydrogen linkages, and crosslinking through chemical bonds. In EPS gels, all these forces can be active, depending on the chemical composition, e.g., proteins vs. polysaccharides. In Table 1, the terminology used in this paper is summarized.

Table 1. Terminology used in this paper.

NOM	natural organic matter
DOM	dissolved organic matter (i.e., passing a filter of about 0.5 μm pore size)
DOC	dissolved organic carbon (i.e., passing a filter of about 0.5 μm pore size)
EPS	exopolymeric substances, found in the colloidal or particulate fraction
TEP	transparent exoplymeric particles, operationally determined
Gels	a type of soft matter that is operationally determined in aquatic systems
HMW	high molecular weight (relative term, usually more than 1 kDa)
LMW	low molecular weight (relative term, usually less than 1 kDa
SFG	surface functional group
DLS	dynamic light scattering
FTIR	fourier transform infrared spectroscopy

Transparent exopolymeric particles, TEP, are commonly considered precursors of EPS [7–9]. They are ubiquitously present in marine and fresh water systems yet 'nonvisible' under the microscope unless they are stained (e.g., Alcian blue; [10]). TEP are primarily assessed as acidic polysaccharides [7]. EPS and TEP do not refer to exactly the same materials: TEP are exopolymers, but not all exopolymeric substances occur as TEP or can form TEP ([7]). EPS forming microbial biofilms have shown to be gels [11–21]. However, TEP are not strictly considered to be gels, as their formation relies on coagulation theory, not on intermolecular energies and assembly processes as for gel formation. Nevertheless, these terms (EPS, TEP, and gels) are often used interchangeably, as in the case of biofilm formation and biofouling [12].

Gels are conceptually considered a type of soft matter [13] and are well-defined [14]. However, microgel concentrations are operationally determined using flow cytometry, after staining with chlortetracycline, expressed as total organic carbon concentration [15], with the kinetics of gel formation determined using dynamic light scattering over hours to days [14]. Gels can also be visualized using environmental electron microscopy [14,16] and/or confocal laser scanning microscopy (e.g., [16]). Coomassie stainable particles (CSP), which are protein-containing particles and can be stained with Coomassie Brilliant Blue, are another type of gel-like particles, proposed by Long and Azam [17], that have been identified in seawater, freshwater, and phytoplankton cultures. TEP and CSP could be discrete particles, or subunits of the same particles [18].

As stated above, EPS are not a defined chemical compound, and their size is in the nano- to micro-size range. TEP are commonly assessed operationally by assaying using the Alcian blue staining of particles collected on a 0.7 μm filter [13–15,20,21]. As has been demonstrated by Hung et al. [22], this method can be biased, but it is nonetheless widely used. EPS are commonly assessed by the sum of the major components, proteins, and polysaccharides of a colloidal or particulate sample [23–25]. Gels, on the other hand, are assessed by flow cytometry, electron microscopy, or dynamic light scattering (DLS) in the filter-passing fraction [14]. Xu et al. [26] were the first to inter-calibrate the three methods, and they found reasonable agreement between them. Before proteins and carbohydrates can be assessed in the filter-passing fraction, EPS have to be pre-concentrated using ultrafiltration, dialysis, or similar techniques. Analytical methods that have been used to determine the major components of EPS include spectrophotometric assays, FTIR, Raman, GC-MS, HPLC, electron microscopy, and NMR. Although proteins and polysaccharides are determined separately, they mostly co-exist in the same macromolecules such as proteoglycans or glycoproteins [9]. Carbohydrates and proteins are determined spectrophotochemically as monomers produced in a sample after a hydrolysis step, and are calibrated against standards, while individual monosaccharides or amino acids can also be determined by HPLC [27]. On the other hand, both polysaccharides and proteins can be more quantitatively determined by NMR and FTIR, as no digestion step is needed [26].

The physico-chemical behavior of EPS (e.g., attachment and aggregation) is mostly determined by the relative hydrophobicity of EPS. Proteins, because of their amphiphilic nature, are considered to contribute most to the relative hydrophobicity of EPS. Their net charge, and thus, their relative hydrophilicity, is dependent on the ambient pH. Amino acids that have hydrophobic side chains are glycine (Gly), alanine (Ala), valine (Val), leucine (Leu), isoleucine (Ile), proline (Pro), phenylalanine (Phe), methionine (Met), and tryptophan (Trp). Individual sugars have different relative hydrophilicities, e.g., pentoses are usually less hydrophilic than hexoses, which is related to the CH-surface area of sugar molecules accessible to water molecules [28].

Proteins are important for the initial attachment process to surfaces [29]. Proteinaceous components of the biofilm matrix include secreted extracellular proteins, cell surface adhesins, and protein subunits of cell appendages such as flagella and pili [30]. Proteins also stabilize the biofilm matrix and three-dimensional biofilm architecture, while proteinaceous enzymes are involved in the degradation of the biofilm components.

The ratio of proteins to carbohydrates of EPS (P/C) has been found to be closely related to the 'stickiness' of EPS and their relative hydrophobicity. For example, the P/C ratio is related to aggregation propensity, e.g., [20], surface tension [21], presence of nano-plastics or oil in microbial cultures [19,31], light-induced chemical crosslinking [23], and, when mineral matter is present, the sedimentation efficiency of marine snow [24]. Figure 1 shows some examples of how these properties can be related to the P/C ratio. Furthermore, the hydraulic residence time or sedimentation efficiency in wastewater treatment systems is also related to the P/C ratio [25]. Protein/carbohydrate ratios of EPS aggregates are thus an indicator of attachment propensity, i.e., its 'stickiness' [32], which can also be directly assessed by magnetic tweezers [33]. Compared with the laborious chemical techniques needed to directly measure protein and carbohydrate content, the P/C ratio can also be expediently obtained with simple fluorescence measurements [31]. The P/C ratio can be a more convenient and informative parameter for the assessment of EPS aggregation behaviors.

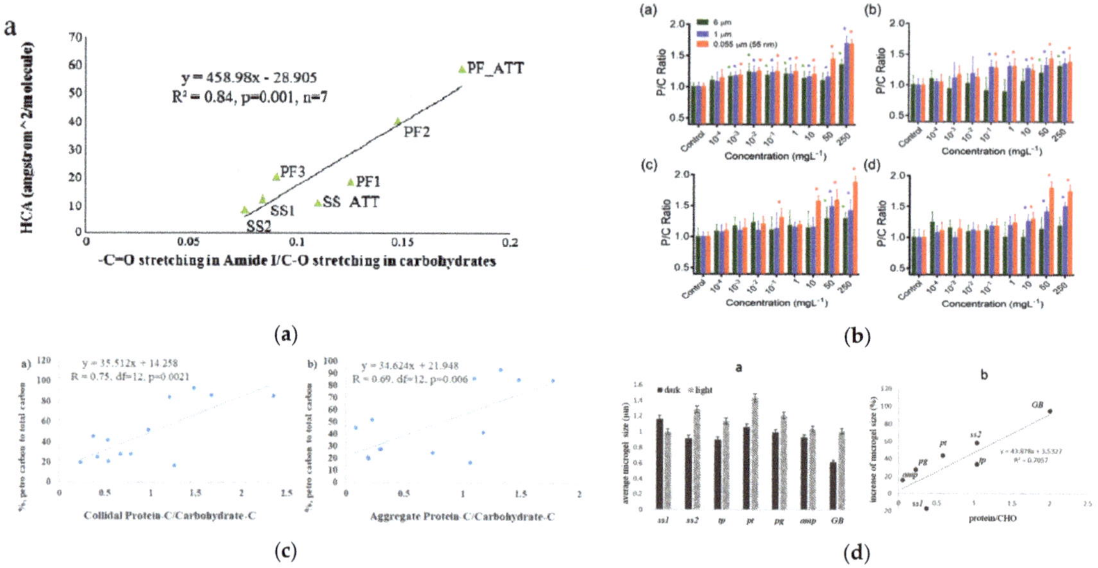

Figure 1. (**a**) Examples for the relative hydrophobicity of EPS that increases with P/C ratio ([20], with permission of the publisher), (**b**) the relationship between nanoplastics concentration and the size-dependent induction of EPS with higher P/C ratio ([19], with permission of the publisher), (**c**) the relationships of % petro-carbon to total carbon in colloidal or sinking aggregates that increase with the P/C ratio of EPS ([34], with permission of the publisher), and (**d**) the microgel size increase due to light-induced ROS chemical crosslinking of proteins in EPS that scale with their P/C ratio ([35], with permission of the publisher).

All these physical properties depend on various physical and chemical factors, such as cross-linker density, cross-liner types, polymer length, pH, types of polymers, temperature, degree of swelling, or temperature. Unfortunately, to the best of our knowledge, there is no available direct measurement for these properties for natural EPS gels in the literature. However, several studies on alginate (or other purified EPS) are available in the literature that might provide some rough assessments. Mechanical and rheological (viscoelastic) properties of alginate gels were shown to be dependent on the cross-linker type, density, ionic conditions, gelling temperatures, or EPS concentrations [36–39]. The specific gravity of a typical synthetic hydrogel (PVA (polyvinyl alcohol) gels) is around 1.05 [40]. For EPS sludges, the density has been reported to be around 1.004–1.048 (g/mL) [41]. For alginate gels depending on the gelling conditions, the value can vary from 1.03 to 1.12 (g/mL) [36]. However, these specific gravity measurements were conducted in non-seawater conditions (in lower salt conditions). Please note that the specific gravity of seawater is around 1.025. The specific gravity of EPS gels in seawater might shift from the measurements in non-seawater environments. The appearance and sizes of marine EPS gels are highly heterogeneous. No typical or characteristic morphology or shape has been found or concluded. The size of EPS gels in seawater can range from sub-micrometers to millimeters, even several centimeters.

EPS are highly heterogeneous mixtures of biopolymers from various microbes in seawater and are generally associated with different types of particles (anthropogenic, minerals, or biological debris), and these physical properties (rheology, morphology, or specific gravity) of natural EPS gels are usually complex and highly variable to determine or generalize.

2. Relative Hydrophobicity of EPS

Exudates from different aquatic organisms can have hydrophilic and hydrophobic moieties. Mostly hydrophilic exudates include the so-called hydrocolloids that are secreted by macro-algae, e.g., seaweeds, and harvested for their distinct chemical properties valued in the food industry as thickening and gelling agents (e.g., [42]). They include acid polysaccharides such as alginates, carrageenans, pectins, gums, and more neutral polysaccharides such as agar and similar substances, extracted from seaweeds, bacteria, and other organisms [42]. Most of these, but not all of them, form gels in the presence of metal ions such as Ca^{2+}. The kind and location of acid functional group determines their food or physiological properties, e.g., alginates are blood coagulants, while carrageenans are anti-coagulants [43].

Due to the solubility limitation of water, hydrophobic moieties of EPS are normally not exposed to the water but, rather, are found in the interior of the structure or proteins or humic substances. As a consequence, EPS and humic substances become amphiphilic. The relative hydrophobicity of these biomolecules, represented by the hydrophobic contact area (HCA), is an important parameter that regulates the kinetics and extent of particle aggregation and dis-aggregation reactions in the water column, and thus influences the removal of associated radionuclides (e.g., Thorium-234) and organic pollutants (e.g., petroleum hydrocarbons). Xu et al. [20] found that the P/C ratio of EPS, determined by FTIR, is linearly related to the HCA, determined by HPLC. This implies that the P/C ratio can be used as an indicator for the relative hydrophobicity of macromolecules. This then also implies that the relative hydrophobicity of the carrier biopolymers of pollutants is mainly controlled by their relative protein contents [34]. High protein content in EPS has also been found to greatly accelerate the formation of marine gels that are not subject to disaggregation after EDTA addition that complexes the Ca^{2+} that are bridging hydrophilic components of EPS, thus rendering gel formation through irreversible hydrophobic interactions [28,44,45].

3. Stability of Microgels upon the Addition of Amphiphiles, e.g., Dispersants

Contrary to the results of [46], which showed the instability of marine gels when irradiated by UV, there is now ample evidence that sunlight irradiation causes reactive oxygen species (ROS)-mediated chemical crosslinking reactions, leading to the photoflocculation of NOM [47]. This was shown through increases in the concentrations of molecular weight, particle size, and mass [23]. On the other hand, global change-induced increases in temperature and hydrogen ion concentrations will have the tendency to decrease the stability of gels [48].

In addition, the aggregation and dispersion of marine gels can be affected by heterogeneous particles and agents in surrounding environments. For example, nano-carbonaceous particles were shown to reduce marine gel formation, due to their negative surface charges interfering with Ca^{2+} bridge cross-linking [49]. This observation is consistent with Zhang et al. [50], which suggests that quantum dots with negative charges have a stronger capability to stabilize EPS gel than positively charged ones. In addition, the microgel size significantly decreased when in the presence of surfactants, especially in the anionic type. Furthermore, negatively charged surfactants such as sodium dodecyl sulfate (SDS) can disrupt existing native microgels, converting larger aggregates into smaller particles. Notably, in addition to human-made pollutants, the input of natural substances can also cause changes in the dynamics of microgels. Shiu et al. [51] indicated that the self-assembly of marine gels would be decreased in the presence of NOM such as Suwannee River humic acid, fulvic acid, and natural organic matter at low concentrations (0.1–10 mg L^{-1}). As mentioned above, a reduction in marine microgel size induced by various specific conditions could lead to a decrease in the downward flux of nutrients and organic carbon, thereby disturbing the organic carbon cycle and biological pump.

Chiu et al. [52] demonstrated that the application of the dispersant Corexit (used to disrupt oil spills) can inhibit EPS aggregation and/or disperse pre-existing microgels in laboratory studies. To represent potential situations during oil spills, a water-accommodated fraction (WAF) of oil and a chemical enhanced WAF (CEWAF) were prepared by mixing oil and dispersant in artificial seawater. It was found that CEWAF can enhance EPS aggregation, with more aggregates accumulating at the air–water interface. While more hydrophobic EPS forms (higher P/C ratio) showed a high resistance to Corexit dispersion, hydrophilic EPS (lower P/C ratio) dispersed more readily when the dispersant Corexit was added, thereby suggesting that P/C ratio plays an important role in determining the stability of microgels in the presence of dispersants. In addition, Shiu et al. [31] showed a negative correlation between P/C ratio and the relative amount of extracellular DNA in EPS, indicating that a higher cellular stress level when exposed to pollutants (WAF and CEWAF) is associated with EPS of higher P/C ratios, resulting in a lower concentration of DNA. This suggests that marine microbes can actively modify their EPS release and composition in response to oil and Corexit treatments.

4. Incorporation of Oil and Other Hydrophobic Pollutants into Gel-Forming EPS and Marine Aggregates

Much has been written on the role of EPS-containing aggregates ('Marine Snow') in accomodating oil and forming 'Marine Oil Snow', MOS [53–57]. Even though EPS gels are normally hydrophilic on the outside, and they hide hydrophobic entities of mostly proteins in their interiors, hydrophobic pollutants such as oil can be accommodated well within gels upon unfolding of the proteins. EPS were found to be crucial to the formation of marine oil snow (MOS), which can form in the presence and absence of Corexit [27]. Using a radiocarbon mass balance or ^{13}C-NMR quantification after a dichloromethane extraction, it was found that the presence of dispersants enhanced the amounts of protein and oil being incorporated into oil-carrying aggregates, yet slowed the sedimentation efficiency of the MOS [24]. EPS with higher P/C ratios (i.e., greater hydrophobicity) tended to facilitate the incorporation of oil and/or Corexit, and the formation of oil-carrying aggregates. When not enough mineral matter is present, colloidal aggregates can become less able to sink due to the lowered density caused by petroleum components. These observations and assessments were confirmed in subsequent mesocosm experiments that simulated both near-shore and off-shore conditions, resulting in significant relationships between the P/C ratio in aggregates/colloids and the percentage of petrocarbon incorporation into these phases regardless of conditions [34]. The P/C ratio of EPS in both the aggregate and the colloidal fraction was thus a key factor for regulating the oil contribution to the sinking aggregates. These studies also pointed out the necessity to consider more closely the presence of a mineral phase, as ballast, to overcome the buoyancy effects of oil in the oil-carrying EPS aggregates.

EPS (as the sum of individually determined proteins and carbohydrates), extracted by EDTA from the surface-attached fraction of particles in mesocosm experiments (with and without oil), correlated well with TEP [26], supporting the use of EPS as a surrogate for TEP measurements in experiments in the presence of Corexit, where TEP cannot be determined, due to analytical interference.

The water solubility of other hydrophobic pollutants such as dioxins, and PAHs, which are normally only sparingly soluble, can be greatly enhanced by their association with 'dissolved' organic carbon (operationally defined as passing through a 0.7 or 0.5 μm filter), which contains natural colloidal, macromolecular organic matter (operationally defined as the fraction retained by an ultrafilter of 1 or 10 kDa pore size, and passing through a 0.5 or 0.7 μm filter) composed of EPS and humic substances ([58]). For example, empirical relationships describing the binding of hydrophobic organic compounds to sedimentary (K_d) and colloidal matter (K_c) have been proposed and experimentally verified. The reader is referred to numerous reviews on this subject, e.g., Schwarzenbach et al. [44]. While this solubility enhancement is important for transport in more turbulent aquatic systems, in water-submerged waste disposal sites, it has been found, using state-of-the art techniques, that the truly dissolved ($\leq$1 kDa fraction) concentration of dioxins in a waste pit was even lower than predicted from K_{ow} and BC values [45].

5. Specific and Nonspecific Interactions of Marine Gels with Metal Ions

The various interactions of metal ions with natural macromolecular organic molecules were reviewed in Buffle [58], Guo et al. [59], Doucet et al. [60], and Santschi et al. [32,61,62]. There are some main differences between trace metal complexation to a low-molecular-weight (LMW) ligand (e.g., citric acid) and to a high-molecular-weight (HMW) polyelectrolyte complexant, whereby the same functional group is attached to either a simple molecule or a macromolecular backbone (e.g., acid polysaccharides, macromolecular thiols, carboxylates, and proteins). LMW ligands have a small number of metal-specific functional groups, while multiple HMW ligands can be attached to different locations in the macromolecule, from where they can chelate trace metals in different ways. The nature of those ligands is relatively well recognized. The interaction between ligands and cations is generally divided into two categories depending on the hardness and softness of acids and bases. Hard acids and bases are characterized by small size, high electronegativity, and low polarizability, including A-type (e.g. Al^{3+}) metals and F, O, and N. They are readily hydrated and tend to form outer-sphere complexes by ionic bonds. Soft acids and bases are characterized by relatively large size, low electronegativity, and high polarizability, including B-type metals (e.g., Ag^{+} and Hg^{2+}) and S, I, and Br. They usually exist dehydrated and tend to form inner-sphere complexes by covalent bonds, which are far more stable than outer-sphere complexes.

High molecular (HMW) ligands are thus macromolecules that have a large number of surface functional groups (SFGs). They are composed of humic substances, polysaccharides, amino acids and peptides, and hydrocarbons. SFGs would be present on the outside of the biopolymer as they present themselves to the water. Due to the more complicated architecture of these biopolymers, the actual architecture can change depending on conditions (e.g., pH, redox, and salinity), and micelles could form at higher concentrations of colloidal forms of NOM. Advanced reverse osmosis/electrodialysis that consistently recovers 68 $\pm$ 2% of DOC allowed the molecular-level characterization of this macromolecular fraction via various spectroscopic (including advanced NMR) techniques [63]. It was found that condensed aromatic and quaternary anomeric carbons contribute to this deep refractory DOC pool, the quaternary anomeric carbons being a newly identified and potentially important component of bio-refractory carbohydrate-like carbon. Their results support the multi-pool (e.g. 3-pool: labile, semi-labile, and refractory) conceptual model of marine DOM biogeochemistry. Therefore, the average values of chemical (stability constants for complexation, acid-base, etc.) or physical properties (e.g., residence times) are sometimes not very meaningful and are subject to biases. The secondary effects that make up such biases can be categorized into several major groups or categories.

(1) Polyfunctional properties: They have various kinds of SFGs (R-COOH, R-OH, R-SH, R-NH_2, etc.). Those different SFGs also have different affinities to hard and soft cations ([44]. Sometimes, a metal ion is bound to more than two SFGs. There may be competition for cations between different SFGs. For example, B-type metals have stronger affinities to (S, S) > (S, N) > (N, O) > (O, O). The same SFG can have different properties depending on the types of backbone (aliphatic or aromatic) to which they bind. Finally, the geometry, such as cavities formed near SFGs, and flexibility of the organic molecules can make a significant difference to the stability of the complexation. These kinds of steric factors are controlled by ionic strength and pH in bulk solution.

(2) Conformational changes: Depending on the hydration/dehydration processes, hydrogen bonds between hydrated cations and SFGs, or metallic bridges, and the conformation of the macromolecules can form aggregates or gels. The hydration water has a different structure from that of water in the bulk solution, and it makes the stability different [58,64]. The nature of a SFG in both LMW ligand and HMW macromolecules is similar. However, the fate of the same SFG may differ depending on the fate of particles and dissolved solutes.

(3) Polyelectric properties: HMW macromolecules have SFGs (e.g., R-COOH) that protonate at low pH and deprotonate at high pH. When they deprotonate under basic

conditions, negatively charged SFGs repulse one another. This process creates an electric field and causes more energy needed to remove protons from SFGs, eventually increasing the pK_a. The formation of electric fields depends on the proportion of protonated sites. This indicates that the degree of protonation or deprotonation is not solely controlled by pH in bulk solution, but also by the near-field interactions between potential ligands.

(4) Binding heterogeneity effects, with binding constants becoming a function of the metal ion-to-surface site ratio [58], occur because the strongest ligands are present at the lowest concentrations, while weaker ligands are present at higher concentrations. This necessitates experimental assessments at ambient concentrations of metals and ligands, or at least use the proper ratios.

(5) 'Particle concentration effects' on kinetic constants (ki) and particle–water partition coefficients (K_d) are a consequence of incomplete separation between particles and solution and colloids, as there often are strong metal complexing macromolecular ligands in the 0.45 μm filter-passing fraction. This effect causes experimentally determined K_d and k_i values to become a function of particle (C_p) concentration. This 'particle concentration effect' on kinetic constants (k_i) and particle–water partition coefficients (K_d) was demonstrated using, as an example, thorium ions in the ocean, that is, the Brownian pumping model of Honeyman and Santschi [65].

Both humic substances [66] and EPS [67] can be considered a heterogeneous Donnan gel phase, similar to the situation in mucus [68]. Donnan equilibria can dominate the exchange of cations and anions across EPS gels surrounding microbial cells. For example, the Donnan mechanism affects mucin release [68] and mucus hydration [69], the swelling of exocytosed polymer-gels in *Phaeocystis pouchetii* [67], and the cation exchange membrane behavior of EPS in salt-adapted granular sludge [70]. Furthermore, toxic effects in saline environments on microbial consortia can be alleviated by the selective binding of cations to negatively charged EPS surrounding their cells, which prevents their diffusion into the deeper parts of the biofilm. The toxic effects of metal cations have been explained by various mechanisms, i.e., their ability to replace metallic enzyme cofactors, thereby disrupting the biological function of these cofactors, and the induction of redox reactions with cellular thiols, provoking Fenton-type reactions that produce reactive oxygen species and by interference with membrane transport processes [70].

When macromolecules form gels, as in the case of EPS, there will be other nonspecific interactions during crosslinking. For example, the crosslinking ability of counter-ions and the stability of the resulting networks increase with the second power of the valence electrons [71]. As an example, Fe^{3+} or Al^{3+} should be able to cross-link dissolved organic matter (DOM) at a fraction of the concentration that Ca^{2+} does. Furthermore, the degree of interaction for trivalent metal ions is higher as compared to that for divalent metal ions at physiological pH (pH $\sim$ 7.0) [72]). Moreover, organic polycations including spermine or spermidine, the condensing peptides of nucleic acids released from dead cells, have four cationic sites and are found in seawater at nanomolar concentrations [73]. They could be very powerful DOM crosslinkers even at submicromolar concentrations. Other polycations such as porins released by bacteria could also be interesting candidates to evaluate.

Felz et al. [74] recently reviewed evidence of how metal ions impact structural EPS hydrogels from aerobic granular sludge. They reported that structural EPS contain alginate hydrogels, but the two are not the same. Structural EPS are more protein-rich, and their gel forming ability, stiffness determined by the Young's modulus, and binding ability are better in the presence of transition metals (rather than alkaline earth metals) than for alginates. They also showed that structural EPS are highly complex, and they have different gelling mechanisms than the acidic polysaccharides alginate, polygalacturonic acid, and kappa carrageenan. In addition, the structural EPS hydrogels show strong integrity toward the chelating reagent EDTA.

It was clearly demonstrated that riverine and marine DOM polymers have the capacity for scavenging selected heavy metals via aggregation processes [75]. The presence of many anionic functional groups on the surface of polymers may provide cation exchange sites for complexing heavy or trace metals [76]. The highest binding affinity in three selected metals was for Cu ions, followed by Ni and Mn ions. The affinity trend is in agreement with the Irving–Williams Series (Cu > Ni > Mn) and other marine colloidal studies. For example, a higher level of colloidal Cu than colloidal Ni was found in coastal areas such as the Danshuei River estuary, Taiwan [77]; the San Francisco Bay estuary, USA [78]; and Narragansett Bay, Rhode Island [79]. This may indicate that gels binding with metals are generally affected by ligand interaction and types of polymers [75].

6. Gel Interactions with Nanoparticles

Studies have shown that the secretion of EPS from microbial cells is significantly affected by surrounding environmental conditions. For example, EPS with higher protein-to-carbohydrate (P/C) ratios are induced by unfavorable growth conditions, including nutrient limitation, toxins (nanoparticles, oil, and dispersant), and light exposure [16,35]. EPS with high P/C ratios are more hydrophobic and sticky, and are thus able to physically block or chemically 'quench' the hazardous agents [80], resulting in the lower effective concentration of toxins entering microbial cells. Furthermore, protein-rich EPS can potentially facilitate faster assembly rates of marine aggregates and alter their aggregation sizesv. The interactions between microbe growth/survival and critical characteristics of EPS (P/C ratio) in the presence of micro- and nano-plastics have received little attention. Therefore, understanding the complex biochemical interactions between three key components (microbes, nanoparticles, and EPS) during nanoparticle (NP) exposure is important to elucidate the fate of NPs, e.g., plastics, especially in their aggregation and scavenging processes in marine environments.

7. Gel Interactions with Micro- and Nano-Plastics

Microplastics, as with many other micro- and nanoparticles, are rapidly covered by biofilms that then further interact with marine biogenic particles [81] to marine plastic snow (MPS). An example of nano-plastic particles enmeshed in EPS and phytoplankton cells, i.e., MPS, is given in Figure 2.

The first step is the formation of microgels. In the study by Ding et al. [15], it was shown that EPS microgel formation in seawater was greatly accelerated by small amounts of amphiphilic EPS or nano-plastics. Later, Chen et al. [16] showed that phytoplankton EPS microgel formation is also accelerated by nano-plastics in seawater, but to a lesser extent. In both cases, it was demonstrated that hydrophobic interactions dominated, which were not affected by EDTA additions, in contrast to hydrophilic interactions. Phytoplankton EPS microgel formation is greatly accelerated by nano-plastics in seawater, likely due to a higher protein content of the EPS produced [19,20]. Furthermore, EPS microgel formation from DOM was greatly accelerated by nano-plastics in different river and lake waters, as well as seawater [19]. Patches of algal cells with 1 μm polystyrene micro-particles encased in an EPS matrix have been observed [19], supporting the hypothesis that plastics would be incoporated into the phytoplankton EPS matrix to form aggregates (marine plastic snow). The sinking route of marine aggregates can scavenge micro- and nano-plastics, which may explain why the negative mass balance of entering vs. measured marine plastics in the surface waters is still lower than expected [15]. EPS effects should be considered in models for predicting and understanding the fate and transport of marine plastic debris.

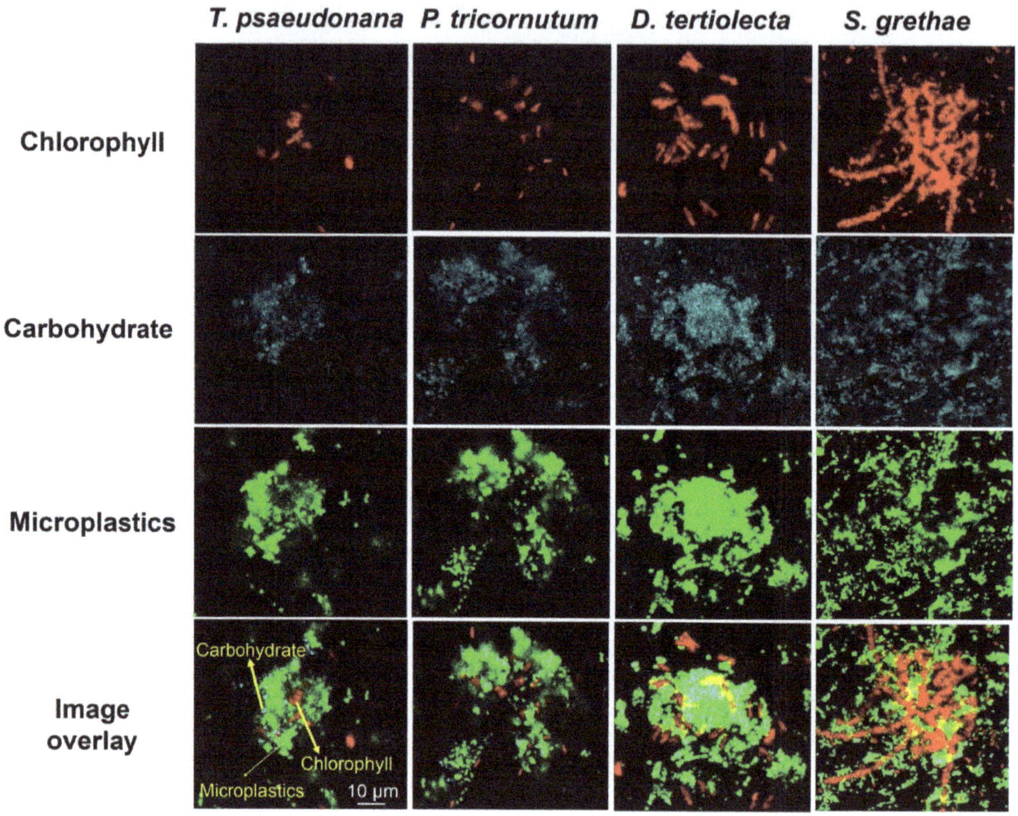

Figure 2. Marine Plastic Snow ([19], with permission from the publisher).

8. Gel Interactions with Organisms

Gel interactions with marine organisms are quite diverse. For example, Haye et al. [15] showed for the first time that filter-feeding organisms effectively filter suspended particles, as well as negatively charged colloidal-sized nanoplastics (0.04 μm), at ambient (1 mg/L) concentrations, most likely as gel-like marine plastic snow. In addition, these authors showed that the chemical composition of EPS controls the uptake of metal ions from the water by oysters. For example, alginate gels greatly ameliorated the metal ion uptake, while carrageenans did not, when compared to ambient colloidal-sized EPS from Galveston Bay. Finally, biofilms composed of EPS gels growing on plastics are reservoirs for antibiotic and metal resistance genes in marine environments. Additionally, marine organisms would be unable to discriminate between target food sources and organic aggregates. These self-assembled microgels concentrate organic and inorganic materials (as above sections), and thus, the accumulation may affect the exposure of higher-trophic-level predators such as zooplankton, invertebrates, filter-feeding fish, and even humans [82,83].

9. Conclusions

Marine gels (analytically determined as gels, EPS or TEP) are ubiquitous, as they form from amphiphilic microbial exudates of macromolecular exopolymeric substances in aquatic systems, and they provide a physical barrier surrounding the microbial cells, mediating the transport of hydrophilic and hydrophobic compounds to and from the cells. They are mostly composed of proteins and polysaccharides, whose relative contribution can change in response to environmental conditions, e.g., the presence of pollutants. These gels, because of their amphiphilic nature, can strongly interact with ionic and nonionic constituents in various ways, e.g., through both hydrophilic and hydrophobic interactions that facilitate certain physical, chemical, or biological pathways.

10. Open Questions

How do marine snow and macrogels of millimeter and centimeter size form from microgels of 5 μm size?—A good start to answer this question has been made by Buffle et al. [84] in their three-colloidal component approach. In their model, they propose that large aggregates form through floc formation by crosslinking many of the individual polysaccharide-rich fibrils. However, more work needs to be carried out to understand the complex physical, chemical, and biological interactions that lead to floc and aggregate formation.

How do marine gels made from EPS respond to changes in environmental conditions, e.g., temperature, pH, redox, nutrient, ionic composition, and strength?—While global-change-related increases in temperature and hydrogen ion concentrations had been addressed by Chen et al. [48], other relevant changes in environmental conditions had not been properly addressed.

Author Contributions: All authors contributed equally to the article. All authors have read and agreed to the published version of the manuscript.

Funding: This research received no external funding.

Conflicts of Interest: The authors declare no conflict of interest.

References

1. Santschi, P.; Balnois, E.; Wilkinson, K.; Zhang, J.; Buffle, J.; Guo, L. Fibrillar polysaccharides in marine macromolecular organic matter as imaged by atomic force microscopy and transmission electron microscopy. *Limnol. Oceanogr.* **1998**, *43*, 896–908. [CrossRef]
2. Droop, M.R. Vitamins, phytoplankton and bacteria: Symbiosis or scavenging? *J. Plankton Res.* **2007**, *29*, 107–113. [CrossRef]
3. Kazamia, E.; Sutak, R.; Paz-Yepes, J.; Dorrell, R.G.; Vieira, F.R.J.; Mach, J.; Morrissey, J.; Leon, S.; Lam, F.; Pelletier, E.; et al. Endocytosis-mediated siderophore uptake as a strategy for Fe acquisition in diatoms. *Sci. Adv.* **2018**, *4*, eaar4536. [CrossRef] [PubMed]
4. Butler, A. Marine Siderophores and Microbial Iron Mobilization. *Biometals* **2005**, *18*, 369–374. [CrossRef] [PubMed]
5. Chin, W.-C.; Orellana, M.V.; Quesada, I.; Verdugo, P. Secretion in Unicellular Marine Phytoplankton: Demonstration of Regulated Exocytosis in Phaeocystis globosa. *Plant Cell Physiol.* **2004**, *45*, 535–542. [CrossRef] [PubMed]
6. Rabin, N.; Zheng, Y.; Opoku-Temeng, C.; Du, Y.; Bonsu, E.; Sintim, H.O. Biofilm formation mechanisms and targets for developing antibiofilm agents. *Future Med. Chem.* **2015**, *7*, 493–512. [CrossRef] [PubMed]
7. Passow, U. Transparent exopolymer particles (TEP) in aquatic environments. *Prog. Oceanogr.* **2002**, *55*, 287–333. [CrossRef]
8. Wurl, O.; Miller, L.; Vagle, S. Production and fate of transparent exopolymer particles in the ocean. *J. Geophys. Res.* **2011**, *116*, C00H13. [CrossRef]
9. Decho, A.W.; Gutierrez, T. Microbial Extracellular Polymeric Substances (EPSs) in Ocean Systems. *Front. Microbiol.* **2017**, *8*, 922. [CrossRef]
10. Passow, U.; Alldredge, A.L. A dye-binding assay for the spectrophotometric measurement of transparent exopolymer particles (TEP). *Limnol. Oceanogr.* **1995**, *40*, 1326–1335. [CrossRef]
11. Bar-Zeev, E.; Berman-Frank, I.; Girshevitz, O.; Berman, T. Revised paradigm of aquatic biofilm formation facilitated by microgel transparent exopolymer particles. *Proc. Natl. Acad. Sci. USA* **2012**, *109*, 9119–9124. [CrossRef] [PubMed]
12. Berman, T. TEP, an Ubiquitous Constituent of NOM is an important factor in Membrane Biofouling. In Proceedings of the 2011 IWA Specialty Conference on Natural Organic Matter, Costa Mesa, CA, USA, 27–29 July 2011.
13. De Gennes, P.G. Soft matter. *Rev. Mod. Phys.* **1992**, *64*, 645–648. [CrossRef]
14. Chin, W.-C.; Orellana, M.V.; Verdugo, P. Spontaneous assembly of marine dissolved organic matter into polymer gels. *Nature* **1998**, *391*, 568–572. [CrossRef]
15. Ding, Y.X.; Chin, W.C.; Rodriguez, A.; Hung, C.C.; Santschi, P.H.; Verdugo, P. Amphiphilic exopolymers from Sagittula stellata induce DOM self-assembly and formation of marine microgels. *Mar. Chem.* **2008**, *112*, 11–19. [CrossRef]
16. Chen, C.-S.; Anaya, J.M.; Zhang, S.; Spurgin, J.; Chuang, C.-Y.; Xu, C.; Miao, A.-J.; Chen, E.Y.T.; Schwehr, K.A.; Jiang, Y.; et al. Effects of Engineered Nanoparticles on the Assembly of Exopolymeric Substances from Phytoplankton. *PLoS ONE* **2011**, *6*, e21865. [CrossRef] [PubMed]
17. Long, R.A.; Azam, F. Abundant protein-containing particles in the sea. *Aquat. Microb. Ecol.* **1996**, *10*, 213–221. [CrossRef]

18. Thornton, D.C.O. Coomassie Stainable Particles (CSP): Protein Containing Exopolymer Particles in the Ocean. *Front. Mar. Sci.* **2018**, *5*. [CrossRef]
19. Shiu, R.-F.; Vazquez, C.I.; Tsai, Y.-T.; Torres, G.V.; Chen, C.-S.; Santschi, P.H.; Quigg, A.; Chin, W.-C. Nano-plastics induce aquatic particulate organic matter (microgels) formation. *Sci. Total Environ.* **2020**. [CrossRef]
20. Xu, C.; Zhang, S.; Chuang, C.-y.; Miller, E.J.; Schwehr, K.A.; Santschi, P.H. Chemical composition and relative hydrophobicity of microbial exopolymeric substances (EPS) isolated by anion exchange chromatography and their actinide-binding affinities. *Mar. Chem.* **2011**, *126*, 27–36. [CrossRef]
21. Schwehr, K.A.; Xu, C.; Chiu, M.-H.; Zhang, S.; Sun, L.; Lin, P.; Beaver, M.; Jackson, C.; Agueda, O.; Bergen, C.; et al. Protein: Polysaccharide ratio in exopolymeric substances controlling the surface tension of seawater in the presence or absence of surrogate Macondo oil with and without Corexit. *Mar. Chem.* **2018**, *206*, 84–92. [CrossRef]
22. Hung, C.; Guo, L.; Santschi, P.; Alvarado-Quiroz, N.; Haye, J. Distributions of carbohydrate species in the Gulf of Mexico. *Mar. Chem.* **2003**, *81*, 119–135. [CrossRef]
23. Sun, L.; Xu, C.; Chin, W.C.; Zhang, S.; Lin, P.; Schwehr, K.A.; Quigg, A.; Chiu, M.-H.; Chin, W.-C.; Santschi, P.H. Light-induced aggregation of microbial exopolymeric substances. *Chemosphere* **2017**, *181*, 675–681. [CrossRef] [PubMed]
24. Xu, C.; Zhang, S.; Beaver, M.; Wozniak, A.; Obeid, W.; Lin, Y.; Wade, T.L.; Schwehr, K.A.; Lin, P.; Sun, L.; et al. Decreased sedimentation efficiency of petro- and non-petro-carbon caused by a dispersant for Macondo surrogate oil in a mesocosm simulating a coastal microbial community. *Mar. Chem.* **2018**, *206*, 34–43. [CrossRef]
25. Ren, B.; Young, B.; Variola, F.; Delatolla, R. Protein to polysaccharide ratio in EPS as an indicator of non-optimized operation of tertiary nitrifying MBBR. *Water Qual. Res. J.* **2016**, *5*, 297–306. [CrossRef]
26. Xu, C.; Chin, W.-C.; Lin, P.; Chen, H.M.; Lin, P.; Chiu, M.-C.; Waggoner, D.C.; Xing, W.; Sun, L.; Schwehr, K.A.; et al. Marine Gels, Extracellular Polymeric Substances (EPS) and Transparent Exopolymeric Particles (TEP) in natural seawater and seawater contaminated with a water accommodated fraction of Macondo oil surrogate. *Mar. Chem.* **2019**, *215*, 103667. [CrossRef]
27. Xu, C.; Zhang, S.; Beaver, M.; Lin, P.; Sun, L.; Doyle, S.M.; Sylvan, J.B.; Wozniak, A.; Hatcher, P.G.; Kaiser, K.; et al. The role of microbially-mediated exopolymeric substances (EPS) in regulating Macondo oil transport in a mesocosm experiment. *Mar. Chem.* **2018**, *206*, 52–61. [CrossRef]
28. Janado, M.; Yano, Y. Hydrophobic nature of sugars as evidenced by their differential affinity for polystyrene gel in aqueous media. *J. Solut. Chem.* **1985**, *14*, 891–902. [CrossRef]
29. Petrova, O.E.; Sauer, K. Sticky situations: Key components that control bacterial surface attachment. *J Bacteriol.* **2016**, *104*, 2413–2425. [CrossRef]
30. Fong, J.N.C.; Yildiz, F.H. Biofilm Matrix Proteins. *Microbiol. Spectr.* **2015**, *3*. [CrossRef]
31. Shiu, R.-F.; Chiu, M.-H.; Vazquez, C.I.; Tsai, Y.-Y.; Le, A.; Kagiri, A.; Xu, C.; Kamalanathan, M.; Bacosa, H.P.; Doyle, S.M.; et al. Protein to carbohydrate (P/C) ratio changes in microbial extracellular polymeric substances induced by oil and Corexit. *Mar. Chem.* **2020**, *223*, 103789. [CrossRef]
32. Santschi, P.H.; Xu, C.; Schwehr, K.A.; Lin, P.; Sun, L.; Chin, W.C.; Kamalanathan, M.; Bacosa, H.P.; Quigg, A. Can the protein/carbohydrate (P/C) ratio of exopolymeric substances (EPS) be used as a proxy for their 'stickiness' and aggregation propensity? *Mar. Chem.* **2020**, *218*, 103734. [CrossRef]
33. Chen, C.-S.; Shiu, R.-F.; Hsieh, Y.-Y.; Xu, C.; Vazquez, C.I.; Cui, Y.; Hsu, I.C.; Quigg, A.; Santschi, P.H.; Chin, W.-C. Stickiness of Extracellular Polymeric Substances on different surfaces via Magnetic Tweezers. *Sci. Total Environ.* **2021**. [CrossRef]
34. Xu, C.; Lin, P.; Zhang, S.; Sun, L.; Xing, W.; Schwehr, K.A.; Chin, W.-C.; Wade, T.L.; Knap, A.H.; Hatcher, P.G.; et al. The interplay of extracellular polymeric substances and oil/Corexit to affect the petroleum incorporation into sinking marine oil snow in four mesocosms. *Sci. Total Environ.* **2019**, *693*, 133626. [CrossRef]
35. Sun, L.; Chin, W.-C.; Chiu, M.-H.; Xu, C.; Lin, P.; Schwehr, K.A.; Quigg, A.; Santschi, P.H. Sunlight induced aggregation of dissolved organic matter: Role of proteins in linking organic carbon and nitrogen cycling in seawater. *Sci. Total Environ.* **2019**, *654*, 872–877. [CrossRef] [PubMed]
36. Jeong, C.; Kim, S.; Lee, C.; Cho, S.; Kim, S.-B. Changes in the physical properties of calcium alginate gel beads under a wide range of gelation temperature conditions. *Foods* **2020**, *9*, 180. [CrossRef] [PubMed]
37. Lee, K.Y.; Rowley, J.A.; Eiselt, P.; Moy, E.M.; Bouhadir, K.H.; Mooney, D.J. Controlling mechanical and swelling properties of alginate hydrogels independently by cross-linker type and cross-linking density. *Macromolecules* **2000**, *33*, 4291–4294. [CrossRef]
38. Lotti, T.; Carretti, E.; Berti, D.; Montis, C.; Del Buffa, S.; Lubello, C.; Feng, C.; Malpei, F. Hydrogels formed by anammox extracellular polymeric substances: Structural and mechanical insights. *Sci. Rep.* **2019**, *9*, 1–9.
39. Kakita, H.; Kamishima, H. Some properties of alginate gels derived from algal sodium alginate. In Proceedings of the Nineteenth International Seaweed Symposium, Kobe, Japan, 26–31 March 2007; pp. 93–99.
40. Sajjan, A.; Banapurmath, N.; Tapaskar, R.; Patil, S.; Kalahal, P.; Shettar, A. Preparation of polymer electrolyte hydrogels using poly (vinyl alcohol) and tetraethylorthosilicate for battery applications. In Proceedings of the IOP Conference Series: Materials Science and Engineering, Karnataka, India, 2–3 March 2018; p. 012078.
41. Feng, C.; Lotti, T.; Canziani, R.; Lin, Y.; Tagliabue, C.; Malpei, F. Extracellular biopolymers recovered as raw biomaterials from waste granular sludge and potential applications: A critical review. *Sci. Total Environ.* **2020**, *753*, 142051. [CrossRef]
42. Saha, D.; Bhattacharya, S. Hydrocolloids as thickening and gelling agents in food: A critical review. *J. Food Sci. Technol.* **2010**, *47*, 587–597. [CrossRef]

43. Kuznetsova, T.A.; Andryukov, B.G.; Besednova, N.N.; Zaporozhets, T.S.; Kalinin, A.V. Marine Algae Polysaccharides as Basis for Wound Dressings, Drug Delivery, and Tissue Engineering: A Review. *J. Mar. Sci. Eng.* **2020**, *8*, 481. [CrossRef]
44. Schwarzenbach, R.P.; Gschwend, P.M.; Imboden, D.M. *Environmental Organic Chemistry*, 2nd ed.; Wiley-Interscience, John Wiley and Sons, Inc.: Hoboken, NJ, USA, 2003.
45. Louchouarn, P.; Seward, S.; Cornelissen, G.; Arp, H.P.H.; Yeager, K.M.; Brinkmeyer, R.; Santschi, P.H. Limited mobility of dioxins near San Jacinto Super Fund site (waste pit) in the Houston Ship Channel, Texas due to high amorphous organic carbon. *Environ. Pollut.* **2018**, *238*, 988–998. [CrossRef]
46. Orellana, M.V.; Verdugo, P. Ultraviolet radiation blocks the organic carbon exchange between the dissolved phase and the gel phase in the ocean. *Limnol. Oceanogr.* **2003**, *48*, 1618–1623.
47. Chen, H.; Abdulla, H.A.N.; Sanders, R.L.; Myneni, S.C.B.; Mopper, K.; Hatcher, P.G. Production of Black Carbon-like and Aliphatic Molecules from Terrestrial Dissolved Organic Matter in the Presence of Sunlight and Iron. *Environ. Sci. Technol. Lett.* **2014**, *1*, 399–404. [CrossRef]
48. Chen, C.-S.; Anaya, J.M.; Chen, E.Y.-T.; Farr, E.; Chin, W.-C. Ocean Warming—Acidification Synergism Undermines Dissolved Organic Matter Assembly. *PLoS ONE* **2015**. [CrossRef]
49. Shiu, R.-F.; Chin, W.-C.; Lee, C.-L. Carbonaceous particles reduce marine microgel formation. *Sci. Rep.* **2014**, *4*, 5856. [CrossRef]
50. Zhang, S.; Jiang, Y.; Chen, C.-S.; Spurgin, J.; Schwehr, K.A.; Quigg, A.; Chin, W.-C.; Santschi, P.H. Aggregation, Dissolution, and Stability of Quantum Dots in Marine Environments: Importance of Extracellular Polymeric Substances. *Environ. Sci. Technol.* **2012**, *46*, 8764–8772. [CrossRef] [PubMed]
51. Shiu, R.-F.; Lee, C.-L.; Chin, W.-C. Reduction in the exchange of coastal dissolved organic matter and microgels by inputs of extra riverine organic matter. *Water Res.* **2018**, *131*, 161–166. [CrossRef]
52. Chiu, M.H.; Vazquez, C.I.; Shiu, R.F.; Le, C.; Sanchez, N.R.; Kagiri, A.; Garcia, C.A.; Nguyen, C.H.; Tsai, S.M.; Zhang, S.; et al. Impact of exposure of crude oil and dispersant (Corexit) on aggregation of extracellular polymeric substances. *Sci. Total Environ.* **2019**, *657*, 1535–1542. [CrossRef] [PubMed]
53. Passow, U.; Ziervogel, K.; Asper, V.; Diercks, A. Marine snow formation in the aftermath of the Deepwater Horizon oil spill in the Gulf of Mexico. *Environ. Res. Lett.* **2012**, *7*, 035301. [CrossRef]
54. Passow, U.; Ziervogel, K. Marine Snow Sedimented Oil Released During the Deepwater Horizon Spill. *Oceanography* **2016**, *29*, 118–125. [CrossRef]
55. Passow, U.; Sweet, J.; Francis, S.; Xu, C.; Dissanayake, A.L.; Lin, Y.Y.; Santschi, P.H.; Quigg, A. Incorporation of oil into diatom aggregates. *Mar. Ecol. Prog. Ser.* **2019**, *612*, 65–86. [CrossRef]
56. Passow, U.; Hetland, R.D. What Happened to All of the Oil? *Oceanography* **2016**, *29*, 88–95. [CrossRef]
57. Engel, A.; Endres, S.; Galgani, L.; Schartau, M. Marvelous Marine Microgels: On the Distribution and Impact of Gel-Like Particles in the Oceanic Water-Column. *Front. Mar. Sci.* **2020**, *7*. [CrossRef]
58. Buffle, J. *Complexation Reactions in Aquatic Systems. An Analytical Approach*; Ellis Horwood Pub.: Chichester, UK, 1988.
59. Guo, L.; Sanschi, P.H. Ultrafiltration and its applications to sampling and characterization of aquatic colloids. In *International Union of Pure and Applied Chemistry (IUPAC) Series on Analytical and Physical Chemistry of Environmental Systems*; John Wiley: Hoboken, NJ, USA, 2007.
60. Doucet, F.J.; Lead, J.R.; Santschi, P.H. Colloid-Trace Element Interactions in Aquatic Systems. In *International Union of Pure and Applied Chemistry (IUPAC) Series on Analytical and Physical Chemistry of Environmental Systems*; John Wiley: Hoboken, NJ, USA, 2007.
61. Santschi, P.H. Marine colloids, agents of the self-cleansing capacity of aquatic systems: Historical perspective and new discoveries. *Mar. Chem.* **2018**, *207*, 124–135. [CrossRef]
62. Santschi, P.; Lenhart, J.; Honeyman, B. Heterogeneous processes affecting trace contaminant distribution in estuaries: The role of natural organic matter. *Mar. Chem.* **1997**, *58*, 99–125. [CrossRef]
63. Helms, J.R.; Mao, J.D.; Chen, H.M.; Perdue, E.M.; Green, N.W.; Hatcher, P.G.; Mopper, K.; Stubbins, A. Spectroscopic characterization of oceanic dissolved organic matter isolated by reverse osmosis coupled with electrodialysis. *Mar. Chem.* **2015**, *177*, 278–287. [CrossRef]
64. Doucet, F.J.; Lead, J.R.; Santschi, P.H. *Colloid-Trace Element Interactions in Aquatic Systems*; John Wiley & Son, Inc.: Hoboken, NJ. USA, 2007.
65. Honeyman, B.; Santschi, P. A Brownian-Pumping Model for Oceanic Trace-Metal Scavenging—Evidence from TH-Isotopes. *J. Mar. Res.* **1989**, *47*, 951–992. [CrossRef]
66. Benedetti, M.F.; van Riemsdijk, W.H.; Koopal, L.K. Humic substances considered as a heterogeneous Donnan gel phase. Environ. Sci. Tech. 30: 1805-1813. Humic substances considered as a heterogeneous Donnan gel phase. *Environ. Sci. Techol.* **1996**, *30*, 1805–1813. [CrossRef]
67. Chin, W.; Orellanam, M.V.; Verdugo, P. Exocytosis in Phaeocystis pouchetii: Donnan mechanism of swelling of exocytosed polymer-gels. *EOS Trans. Am. Geophys. Un.* **1996**, *76*, 75.
68. Aitken, L.M.; Verdugo, P. Donnan mechanism of mucin release and conditioning in goblet cells: The role of polygons. *J. Exp. Biol.* **1989**, *53*, 73–79.
69. Tam, P.Y.; Verdugo, P. Control of mucus hydration as a Donnan equilibrium process. *Nature* **1981**, *292*, 340–342. [CrossRef]

70. Sudmalisa, D.; Mubita, T.M.; Gagliano, M.C.; Dinis, E.; Zeeman, G.; Rijnaarts, H.H.M.; Temmink, H. Cation exchange membrane behaviour of extracellular polymericsubstances (EPS) in salt adapted granular sludge. *Water Res.* **2020**, *178*, 115855. [CrossRef]
71. Tanaka, T. Gels. *Sci. Am.* **1981**, *244*, 124–138. [CrossRef]
72. Kanti De, S.; Kanwa, N.; Chakraborty, A. Influence of Trivalent Metal Ions on Lipid Vesicles: Gelation and Fusion Phenomena. *Langmuir* **2019**, *35*, 6429–6440. [CrossRef]
73. Nishibori, N.; Yuasa, A.; Sakai, M.; Fujihara, S.; Nishio, S. Free polyamine concentrations in coastal seawater during phytoplankton bloom. *Fish. Sci.* **2001**, *67*, 79–83. [CrossRef]
74. Felz, S.; Kleikamp, H.; Zlopasa, J.; van Loosdrecht, M.C.M.; Lin, Y. Impact of metal ions on structural EPS hydrogels from aerobic granular sludge. *Biofilm* **2020**, *2*, 100011. [CrossRef] [PubMed]
75. Shiu, R.-F.; Lee, C.-L. Role of microgel formation in scavenging of chromophoric dissolved organic matter and heavy metals in a river-sea system. *J. Hazard. Mater.* **2017**, *328*, 12–20. [CrossRef]
76. Joshi, P.M.; Juwarkar, A.A. In Vivo Studies to Elucidate the Role of Extracellular Polymeric Substances from Azotobacter in Immobilization of Heavy Metals. *Environ. Sci. Technol.* **2009**, *43*, 5884–5889. [CrossRef] [PubMed]
77. Jiann, K.-T.; Wen, L.-S.; Santschi, P.H. Trace metal (Cd, Cu, Ni and Pb) partitioning, affinities and removal in the Danshuei River estuary, a macro-tidal, temporally anoxic estuary in Taiwan. *Mar. Chem.* **2005**, *96*, 293–313. [CrossRef]
78. San˜udo-Wilhelmy, S.A.; Rivera-Duarte, I.; Russell Flegal, A. Distribution of colloidal trace metals in the San Francisco Bay estuary. *Geochim. Cosmochim. Acta* **1996**, *60*, 4933–4944. [CrossRef]
79. Wells, M.L.; Smith, G.J.; Bruland, K.W. The distribution of colloidal and particulate bioactive metals in Narragansett Bay, RI. *Mar. Chem.* **2000**, *71*, 143–163. [CrossRef]
80. Nichols, C.A.M.; Guezennec, J.; Bowman, J.P. Bacterial Exopolysaccharides from Extreme Marine Environments with Special Consideration of the Southern Ocean, Sea Ice, and Deep-Sea Hydrothermal Vents: A Review. *Mar. Biotechnol.* **2005**, *7*, 253–271. [CrossRef] [PubMed]
81. Michels, J.; Stippkugel, A.; Lenz, M.; Wirtz, K.; Engel, A. Rapid aggregation of biofilm-covered microplastics with marine biogenic particles. *Proc. R. Soc. B* **2018**, *285*. [CrossRef] [PubMed]
82. Lusher, A.L.; Hernandez-Milian, G.; O'Brien, J.; Berrow, S.; O'Connor, I.; Officer, R. Microplastic and macroplastic ingestion by a deep diving, oceanic cetacean: The True's beaked whale Mesoplodon mirus. *Environ. Pollut.* **2015**, *199*, 185–191. [CrossRef]
83. Cole, M.; Lindeque, P.; Fileman, E.; Halsband, C.; Goodhead, R.; Moger, J.; Galloway, T.S. Microplastic Ingestion by Zooplankton. *Environ. Sci. Technol.* **2013**, *47*, 6646–6655. [CrossRef]
84. Buffle, J.; Wilkinson, K.J.; Stoll, S.; Filella, M.; Zhang, J. A Generalized Description of Aquatic Colloidal Interactions: The Three-colloidal Component Approach. *Environ. Sci. Technol.* **1998**, *32*, 2887–2899. [CrossRef]

gels

MDPI

Review

Transparent Exopolymer Particles in Deep Oceans: Synthesis and Future Challenges

Toshi Nagata [1,*], Yosuke Yamada [2] and Hideki Fukuda [3]

1 Atmosphere and Ocean Research Institute, The University of Tokyo, Kashiwa 277-8564, Japan
2 Marine Biophysics Unit, Okinawa Institute of Science and Technology Graduate University, Onna-son 904-0495, Japan; yosuke.yamada@oist.jp
3 International Coastal Research Center, Atmosphere and Ocean Research Institute, The University of Tokyo, Otsuchi 028-1102, Japan; hfukuda@aori.u-tokyo.ac.jp
* Correspondence: nagata@aori.u-tokyo.ac.jp

Abstract: Transparent exopolymer particles (TEP) are a class of abundant gel-like particles that are omnipresent in seawater. While versatile roles of TEP in the regulation of carbon cycles have been studied extensively over the past three decades, investigators have only recently begun to find intriguing features of TEP distribution and processes in deep waters. The emergence of new research reflects the growing attention to ecological and biogeochemical processes in deep oceans, where large quantities of organic carbon are stored and processed. Here, we review recent research concerning the role of TEP in deep oceans. We discuss: (1) critical features in TEP distribution patterns, (2) TEP sources and sinks, and (3) contributions of TEP to the organic carbon inventory. We conclude that gaining a better understanding of TEP-mediated carbon cycling requires the effective application of gel theory and particle coagulation models for deep water settings. To achieve this goal, we need a better recognition and determination of the quantities, turnover, transport, chemical properties, and microbial processing of TEP.

Keywords: transparent exopolymer particles; ocean carbon cycles; deep oceans

Citation: Nagata, T.; Yamada, Y.; Fukuda, H. Transparent Exopolymer Particles in Deep Oceans: Synthesis and Future Challenges. *Gels* **2021**, *7*, 75. https://doi.org/10.3390/gels7030075

Academic Editor: Pedro Verdugo

Received: 27 May 2021
Accepted: 19 June 2021
Published: 22 June 2021

1. Introduction

Seawater carries a wide variety of dissolved and particulate organic carbon (DOC and POC, respectively), covering a size range of less than a nanometer (dissolved organic molecules and colloids) to meters (whales), with a broad range of turnover times from minutes to millennia [1,2]. Understanding the magnitude and the spatiotemporal patterns of the production, transformation, and remineralization of organic matter is important not only for studies examining ocean carbon cycles but also for those investigating ocean ecosystems and Earth's climate. One emerging concept in this research field is that gel-like particles play an important role in the regulation of organic carbon dynamics in the oceans. Gel-like particles are omnipresent in marine environments, being produced by microbes and larger organisms, released from decayed cells and tissues, or formed through the spontaneous self-assembly of DOC and subsequent coagulation of small particles [2–4]. Transparent exopolymer particles (TEP), consisting of acidic polysaccharides, are among the most abundant class of gel-like particles in marine environments [5,6]. TEP are mainly produced by phytoplankton and bacteria in the oceans [6]. Due to their sticky nature, TEP facilitate particle coagulation and may enhance the vertical transport of organic carbon and the carbon sequestration in the ocean [7]. TEP are porous and less dense, allowing them to accumulate in the sea surface microlayer, where they influence the air–sea exchange of climate-related gas [5]. TEP may also be a key component of marine food webs, providing habitats and organic substrates for microbes [8,9] and serving as a food for metazoans [10].

Although these processes and dynamics of TEP in marine systems have been extensively studied during the past three decades (reviewed by Passow et al. [6] and Mari et al. [5]), investigators have only recently begun to recognize intriguing features of TEP distribution and processes involved in the regulation of TEP dynamics in deep oceans. The emergence of new research reflects the growing attention to ecological and biogeochemical processes in deep oceans, where large quantities of organic carbon are stored and processed, yet mechanisms underlying these processes are not entirely clear (see reviews [11–14]). Here, we review recent research on TEP in deep ocean realms. We first provide an overview of recent TEP distribution data, which is followed by a discussion concerning sources and sinks of TEP. We then examine the contribution of TEP to the organic carbon pool. Our goal is to identify major knowledge gaps and future research challenges.

2. Overview of the Data on TEP Distribution in Deep Waters

In this paper, the mesopelagic layer refers to the depth layers between 200 and 1000 m, and the bathypelagic layer between 1000 m and the abyssal seafloor (up to the depth of 5400 m including a part of the abyssopelagic layer), unless different boundary depths are used in the source literature.

2.1. Data Obtained by the Colorimetry

The routine method to determine TEP concentration in seawater is colorimetry [15]. Briefly, TEP are collected on 0.4-μm-pore-size polycarbonate filters by filtration and are stained with Alcian blue, a cationic dye that binds to anionic carboxyl or half ester-sulfate groups of acidic polysaccarhides at low pH. After a short period of staining, filters are rinsed with pure water and soaked in acid. The redissolved dye concentration is determined colorimetrically. The TEP concentration is expressed as an equivalent amount of the standard substance, xanthan gum (a polysaccharide excreted by a bacterium, *Xanthomonas campestris*), with units of μg Xeq. L^{-1}. The method measures the amount of dye bound to particles, which is converted to the "mass" using the binding capacity of the standard substance. The mass of TEP determined in this manner may deviate from the "true mass", depending on the anionic density of TEP in natural seawater [16]. Thus, it is important to keep in mind that the colorimetric method is semiquantitative.

2.1.1. Data Collected in Coastal, Slope Region, and Marginal Seas

An early study conducted by Passow and Alldredge [15] examined the depth profile of TEP in Santa Barbara Chanell in the eastern North Pacific, in summer. Their data revealed a low TEP concentration (ca. 20 μg Xeq. L^{-1}), with little variation in concentration between the depths of 200 m and 1400 m. Substantially higher TEP concentrations in the mesopelagic layer (300 and 1000 m) were found by Bar-Zeev et al. [17] during transect cruises conducted in the oligotrophic eastern Mediterranean Sea. They found that TEP concentrations at seven sampling stations were, on average, 200 μg Xeq. L^{-1} at depths of both 300 and 1000 m. A notable feature was that the TEP concentration in the mesopelagic layer tended to decrease with increasing distance from the shore. This off-shoreward decreasing trend in TEP concentration was also observed in the near-surface layer.

Ortega-Retuerta et al. [18] examined the TEP distribution at 29 stations along the east–west transect across the Mediterranean Sea and the adjacent North Atlantic. They found that the TEP concentrations in the meso- and bathypelagic layers were 1.2–35 and 0.6–16 μg Xeq. L^{-1}, respectively. The TEP concentration tended to decrease with depth, with a slope variation among the stations. The depth-integrated TEP values in the epipelagic and the meso- and bathypelagic waters were significantly positively correlated. Based on these results, the authors suggested that the TEP concentrations at depth are likely to be controlled by the vertical delivery of TEP.

Yamada et al. [19] investigated TEP concentrations at eight sampling stations in the slope region of the western Arctic Ocean. They found that the TEP concentration range in the layer between 200 and 1960 m was 37–129 μg Xeq. L^{-1}, which displayed a decreasing tendency with depth. They suggested that the large amount of TEP produced in the Chukchi shelf are laterally transported to the slope region [20], where they then transfer to deeper layers due to the vertical transport of TEP associated with sinking particles and the subsequent dissociation from sinking particles during their transit through deep water columns.

In an estuarine environment (the lower St. Lawrence Estuary in Canada) with a maximum depth of 340 m, Annane et al. [21] examined the vertical distribution of TEP. They found a seasonal variation in the TEP concentration in the mesopelagic layer (130–320 m) over a range from 15 to 200 μg Xeq. L^{-1}, with the highest concentration being observed in spring.

2.1.2. Data Collected in Open Oceans

Cisternas-Novoa et al. [22] examined the full-depth distribution of TEP concentrations at the Bermuda Rise, the open ocean domain of the North Atlantic Ocean, where the maximum depth is about 4500 m. Seawater samples were collected on five occasions covering a seasonal cycle (February, May, August, and November). TEP concentrations below a depth of 200 m were uniformly low (mostly in the range of 20–30 μg Xeq. L^{-1}) and seasonally less variable, with a notable exception found in February. In this month, they observed a TEP concentration peak at a depth of 2000 m. In May and June, a strong benthic nepheloid layer developed at depths below 4000 m. The benthic nepheloid layer is characterized by high turbidity and high particulate matter concentration near the seafloor, which is caused by particle erosion/resuspension and inhibited particle settling due to the bottom current and bottom boundary layer turbulent mixing [23]. The authors' results showed that particle concentration and Coomassie stainable particle (CSP) concentration were indeed high in the benthic nepheloid layer, whereas such a trend was not evident for TEP concentrations.

At three sampling stations located in the subtropical and equatorial regions of the central Pacific, Yamada et al. [19] examined full depth distributions of TEP concentrations. Similar to the results obtained by Cisternas-Novoa et al. [22] in the Atlantic Ocean, the TEP concentration was uniform below a depth of 200 m down to the maximum depth of about 5400 m (range, 12–40 μg Xeq. L^{-1}) (Figure 1). The vertical TEP distribution was largely decoupled from those of prokaryote abundance and production, which decreased by approximately 10-fold (abundance) or 100-fold (production) within the corresponding depth range (Figure 1).

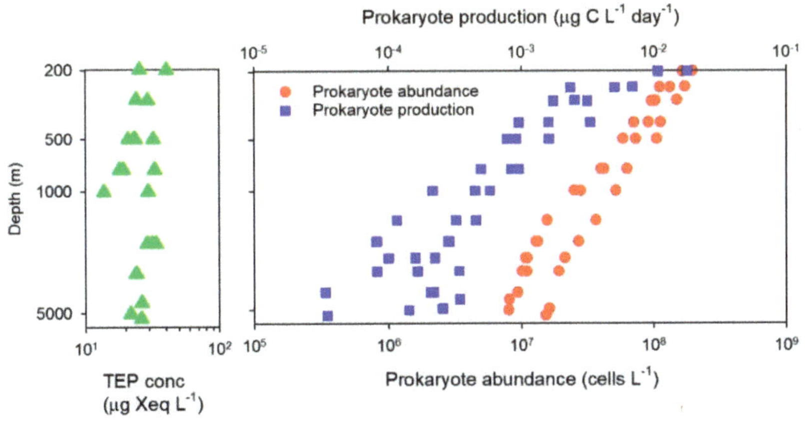

Figure 1. Vertical distributions of TEP concentrations, prokaryote abundance, and prokaryote production in the central Pacific Ocean. The graphs were made using the original data of Yamada et al. [19]. Both x and y axes are logarithmic.

2.2. Data Obtained by the Microscopy

The microscopic method relies on the binding of Alcian blue to TEP [8]. Seawater samples are filtered through 0.4- or 0.2-µm-pore-size polycarbonate filters, stained with Alcian blue, and observed under the light microscope. The size and abundance of TEP are determined either manually or with the aid of an image analysis system, after capturing TEP images with a camera. Because TEP images show the cross-sections of particles on a plane, the area of individual TEP can be determined to calculate the equivalent spherical diameter (ESD) and volume. Using ×200 magnification light microscopy and an image analysis system, the ESD range determined by the microscopy is typically between 1 µm and a few hundred µm, although the size range may vary depending on the study. The microscopic method is time-consuming, even with the aid of an image analysis system; only a few studies have used the microscopic method to analyze TEP in deep oceanic waters [17,24,25]. Although there has been an attempt to use an automated image acquisition system combined with a flow chamber (FlowCAM) to determine TEP abundance and size distribution in a rapid manner, this method has yet to be applied to the analysis of TEP in deep waters [22].

Engel et al. [25] reported the first comprehensive data of TEP abundance determined using a microscopic method in the meso- and bathypelagic oceans. They sampled at six vastly distant oceanic provinces, including subtropical/equatorial, coastal upwelling, and polar regions. Notably, an up to 40-fold higher TEP abundance (in terms of area per unit volume) was found in the Mauritanian upwelling region relative to other regions; moreover, this large difference held for both the meso- and bathypelagic layers. The TEP abundance in the mesopelagic layer was 1.3–3.3-fold higher than the TEP abundance in the bathypelagic layer in the regions examined, except for the subtropical/equatorial region of the Indian Ocean, where TEP abundance in the bathypelagic layer was about 2-fold higher than that in the mesopelagic layer.

Engel et al. [25] also examined the size distribution of TEP (size range, 1–500 µm). They found that the data did not fit well to the general power law model [26], irrespective of sampling depth. This was primarily because the size distribution slope within the size range of 1–6 µm was smaller than that within a larger size range. The relative abundance of large TEP tended to increase with TEP abundance. In the Mauritanian upwelling region where the TEP abundance was high, the size distribution of TEP was skewed toward larger size classes in both the meso- and bathypelagic layers.

In the Fram Strait in the Arctic Ocean, Busch et al. [24] examined the prokaryote colonization of TEP. They found that TEP was colonized by prokaryotes throughout the water column (the maximum depth, 2613 m). Interestingly, the highest density of prokaryotes attached to TEP (58×10^4 cells mm^{-2}) was found at a depth of 1000 m. Prokaryotes attached to TEP accounted for 1–20% (average, 3%) of the total prokaryote abundance. This value tended to increase with depth, with the highest value being found at depths >2000 m.

In the Mediterranean Sea, Bar-Zeev et al. [17] found that large TEP with amorphous shapes were abundant in the mesopelagic layer. The large TEP were often associated with prokaryotes, suggesting that TEP may serve as an organic carbon substrate for prokaryotes in the mesopelagic layer. At the entrance of the Bay of Villefranche (NW Mediterranean Sea), Weinbauer et al. [27] examined TEP seasonal variability at a depth of 300 m. They found that the TEP volume concentration varied within a range of 0.1–0.6 ppm, which increased with temperature.

2.3. Summary of the Observed Data

- In the mesopelagic layer, the variability range of the TEP concentrations is on the order of 100-fold (Table 1). In the marginal sea and slope region, a vertical (depth-dependent decrease [18,19]) and a lateral (offshoreward decrease [17]) gradient in the TEP distribution pattern has been documented. Microscopic observations have revealed that TEP are colonized by prokaryotes in the mesopelagic layer [17,24].

- Examination of the full-depth distribution of TEP in open oceans has revealed that TEP concentrations are less variable (<3 fold) throughout the meso- and bathypelagic water columns down to the depths of 4000–5400 m (Table 1), although some anomalous features have been noted [22,25]. This vertical distribution of TEP is largely decoupled from the distribution of prokaryote abundance and production [19] (Figure 1). One study using the microscopic method has found a remarkably high TEP abundance in the bathypelagic layer of the coastal upwelling region [25]. Microscopic observations have also found that TEP were colonized by prokaryotes in the bathypelagic layer. High relative contributions of TEP-associated prokaryotes to the total prokaryote abundance (up to 20%) were observed at the depths >2000 m in the Arctic Ocean [24].

Table 1. TEP concentrations (μg Xeq. L^{-1}) in the meso- and bathypelagic layers. Values in parentheses are the depth range in meters.

Region	Mesopelagic		Bathypelagic		References
Coastal and slope region, estuary and marginal sea					
Santa Barbara Chanell (eastern Pacific)	20	(200–1400)			[15]
Eastern Mediterranean Sea	200	(300–1000)			[17]
Mediterranean Sea and Atlantic	1.2–35	(200–1000)	0.6–16	(1000–3900)	[18]
Western Arctic (slope region) [1]	37–129	(200–1000)	39–52	(1230–1960)	[19]
St Lawrence Estuary	15–200	(130–320)			[21]
Open oceans					
North Atlantic Ocean (subtropical) [2]	18–33	(200–1000)	16–48	(1250–4580)	[22]
Central Pacific (subtropical and equatorial) [1]	12–40	(200–1000)	14–34	(1000–5370)	[19]

[1] Values from the original data of Yamada et al. [19]; [2] Values extracted from Figure 13 of Cisternas-Novoa et al. [22].

3. Potential Factors Affecting TEP Distribution in the Deep Oceans

As a basis for examining possible mechanisms underlying the observed TEP distribution, here we discuss sources (transport and autochthonous production) and sinks (microbial degradation and grazing) of TEP in deep waters.

3.1. Sources of TEP

3.1.1. Transport

Individual TEP generally sink only slowly, or even float, due to their low density and high porosity [5]. However, if they are incorporated into or associated with dense sinking particles, TEP can be transferred to the deeper layers by gravitational settling. TEP are likely to be released (disaggregated) from sinking particles during their transit in the meso- and bathypelagic water columns. The hydrolysis of the TEP and other polymer networks by ectoenzymes produced by colonizing prokaryotes may enhance the fragmentation of the sinking particles [28]; however, physical processes (turbulence and sheer) and the disturbance caused by zooplankton may also promote fragmentation [26,29].

In oceans, the vertical POC flux (F, mg C m^{-2} d^{-1}) attenuates with depth (Z m). The most common model describing this attenuation is the power law: $F = (F_{100}/100) \times Z^{-0.858}$, where F_{100} is the POC flux at a depth of 100 m [30]. The equation indicates that, of the POC removed from the upper layer, 14% reach a depth of 1000 m, and only 4% are found at a depth of 5000 m. This substantial attenuation of the POC flux with depth is a consequence of fragmentation and remineralization of POC during the transit of particles through the water column [30,31]. Although no previous work has examined the vertical flux attenuation of TEP, the data obtained by Passow et al. [32] support the notion that the depth-dependent attenuation of TEP flux is high. Using time-series sediment traps deployed at a depth of 500 m in the Santa Barbara Channel, they demonstrated that the daily recovery of TEP from sediment traps relative to the TEP standing stocks in the upper water column (0–75 m) was less than 2% (mostly, <0.5%). These values were comparable to those of POC. Therefore, the extent of vertical delivery of TEP associated with settling particles appears to decrease rapidly with depth.

As already mentioned, Yamada et al. [19] found that the TEP concentration decreased with depth in the Arctic slope region. Similarly in the Mediterranean Sea, TEP concentration tended to decrease with depth in the meso- and bathypelagic layers [18]. These depth-dependent trends were interpreted as an indication that the TEP distribution is shaped by the vertical delivery mediated by sinking particles. However, in the open ocean domains of the Pacific and Atlantic Oceans, the depth-dependent variability in TEP concentration was small [19,22] (Figure 1). These apparently contradictory results suggest that the coupling of TEP vertical distribution and sinking POC flux attenuation differs, depending on the environment, such that it is strong in marginal seas and slope regions but weak in open oceans. One could then hypothesize that the vertical delivery of TEP is enhanced by ballast particles (e.g., dense mineral grains) supplied from land and continental shelves and slopes, leading to a magnified effect of sinking particles on TEP vertical distributions in marginal seas and slope regions.

Particle settlement is not the sole physical mechanism by which POC (including TEP) is delivered to deeper waters. Non-sinking particles can be delivered to deeper layers via mixing, convection, and lateral transport along the slope of isopycnal surfaces [33]. The TEP distribution at a particular depth may also be influenced by the transport driven by intermediate water intrusions and deep water currents. In this regard, the anomalously higher TEP concentration and abundance found in the bathypelagic layer relative to the mesopelagic layer is intriguing [22,25]. We clearly need more data concerning the variability in TEP concentrations across different water masses in the meso- and bathypelagic oceans.

3.1.2. Autochthonous Production of TEP

Potential TEP producers in deep oceans are bacteria and archaea (prokaryotes) [6]. Based on the results from a study conducted in the Mediterranean Sea, Ortega-Retuerta et al. [18] suggested that prokaryotes release TEP during their growth. They examined the changes of TEP concentrations in seawater cultures (filtered seawater amended with in-situ prokaryote communities) prepared using meso- and bathypelagic waters. During the 6-day incubation period, they found that the TEP concentration increased with prokaryote abundance. Additionally, the TEP concentration was positively correlated with prokaryote abundance in the meso- and bathypelagic layers. Based on the data, the authors suggested that prokaryotes played a role as a source of TEP in the deep Mediterranean Sea.

The self-assembly of TEP precursors in seawater is another potential autochthonous mechanism of TEP production. Gels are thought to be produced by the spontaneous assembly of polymers to form nanogels in seawater and may contribute to the DOC-to-POC transition [2–4]. Nanogels become larger due to annealing and subsequent coagulation [4,26]. Because high-molecular-weight DOC, containing polymeric substances, is distributed throughout the deep water column [34], the self-assembly of polymeric precursors may explain the uniform TEP vertical distribution observed in the deep waters of the Pacific [19] and Atlantic Oceans [22]. In support of this hypothesis, Ding et al. [35] found that gels were formed by polymer self-assembly in deeper waters of the subtropical Pacific. The concentration of self-assembled gel was similar at depths of 500 and 4000 m. However, currently, there is no evidence that the gels detected by the Ca^{2+} binding assay used by Ding et al. [35] were TEP. According to the gel theory, counterion interactions take place between Ca^{2+} and marine biopolymers, leading to gel production in the ocean [3]. To resolve a critical question concerning whether this theory explains TEP production in deep oceans, further studies are required to eliminate inherent ambiguities of the results obtained by the Alcian Blue assay.

3.2. *Sinks of TEP*

3.2.1. Prokaryotes

In the previous section, we discussed the possibility that prokaryotes produce TEP. However, prokaryotes may also act as decomposers of TEP, whereby they are regarded as a sink of TEP. Several studies have demonstrated that prokaryotes in the meso- and bathypelagic waters express a wide range of ectoenzymatic activities (including phosphatase, beta-glucosidase, and peptidases) that cleave polymeric chains [13]. It is also generally known that the genes coding prokaryotic enzymes that cleave sulfate residues (sulfatase) and carboxyl residues (carboxylase) are widespread and expressed in marine microbes [36]. However, a recent study suggested that a fucose-containing sulphated polysaccharide excreted by diatoms is less susceptible to enzymatic degradation relative to non-sulphated polysaccharide [37]. Therefore, a question arises regarding the lability of TEP in deep waters: Are they good substrates for prokaryotes or not?

Bulk DOC in seawater, especially in deep water, turns over slowly, with the average lifetime being on the order of millennia [1,38]. Although the mechanisms controlling the persistence of DOC have been a subject of much debate, one theory suggests that the inherently recalcitrant molecular properties of DOC explain its resistance to microbial attack [39,40]. Thus, if TEP in deep waters are primarily formed via the spontaneous assembly of DOC [2,4], one would expect that TEP in deep waters are recalcitrant (unavailable for prokaryote consumption). Alternatively, if TEP are formed from or enriched by labile constituents presumably derived from sinking particles [41], TEP may act as "hot spots" for prokaryote substrate consumption. Although we have no conclusive answer regarding this question, the observations of Bar-Zeev et al. [17] and Busch et al. [24] that TEP at depth were colonized by prokaryotes appear to support the "hot spot" hypothesis. Future studies examining the community compositions and gene expression of prokaryotes associated with TEP in deep water may provide insight into this topic. However, for rigorous testing of this hypothesis, we clearly need data about the metabolic activities of the TEP-associated prokaryotes. Currently, no standard method is available to determine the TEP consumption rate by microbes in seawater, making it difficult to evaluate quantitatively the role of prokaryotes as a sink of TEP. Efforts to determine TEP consumption by deep water prokaryotes are further complicated by challenges in evaluating the effects of high hydrostatic pressure on microbial physiology [42,43].

3.2.2. Grazers

Very little is known about the grazing of TEP by metazoan and protist grazers in deep waters. Given that TEP in the meso- and bathypelagic layers are colonized by prokaryotes and that prokaryotes are protein-rich constituents with a low carbon-to-nitrogen ratio [44], the TEP–prokaryotes complex may serve as a good food source for detritivorous grazers. The search for TEP–prokaryotes complexes by metazoan grazers may be facilitated by bacterial bioluminescence [45]. The possibility of the food web control of TEP in the dark ocean deserves further investigation.

4. Organic Carbon Inventory

Yamada et al. [19] estimated the organic carbon concentration of TEP (TEP-C concentration) in the meso- and bathypelagic waters. They found that TEP-C represents a substantial fraction of the POC inventory. In the mesopelagic layer, TEP-C concentrations were 10–23 μg C L^{-1}, accounting for 230% and 320% of the POC in the Arctic and Pacific Oceans, respectively. The corresponding values in the bathypelagic layer of the Pacific Ocean were 41 μg C L^{-1} and 550%. Similarly high-to-moderate contributions of TEP-C to POC in the meso- and bathypelagic layers have been noted in other studies [17,18,21]. Hypotheses to explain the fact that the concentration of TEP-C exceeds POC are listed below.

1. To estimate TEP-C, studies have used a conversion factor derived from laboratory experiments using diatom cultures [15]. However, the validity of this conversion factor in deep waters has yet to be tested. If the organic carbon yield relative to the Alcian blue-reactive residues (sulfate and carboxyl groups) of TEP is systematically lower in deeper than shallower waters, the TEP-C values estimated from the conversion factor for the shallower water (diatom-derived fresh TEP) may be too high.
2. TEP-C concentration may exceed POC concentration due to the use of different pore-size-filters for the determination of TEP (0.4-µm-pore-size polycarbonate filter) and

POC (0.7-µm-pore-size GF/F filter). If large quantities of organic carbon associated with TEP pass through the GF/F filters, but are retained on 0.4-µm polycarbonate filters, this would explain the high TEP-C concentration relative to POC.

Engel et al. [25] estimated the TEP-C concentration using the microscopic data and an equation that relates the size (ESD) of TEP to the organic carbon concentration [44]. TEP-C in the meso- and bathypelagic layers at the six stations examined were in the range of 0.6–1.7 µg C L^{-1} and 0.4–3 µg C L^{-1}, respectively, except for higher values found in the Mauritanian upwelling region where the corresponding values were 48 and 22 µg C L^{-1}, respectively. These TEP-C values, except for those from the Mauritanian upwelling, were about one order of magnitude lower than those reported by Yamada et al. [19]. The discrepancies between the two studies may be explained by regional and seasonal differences, uncertainties associated with the TEP-to-carbon conversion factor or equation, and the difference in the lower size limit of the TEP determination, which was 0.4 µm in Yamada et al. [19] and 1 µm in Engel et al. [25].

Figure 2 compares TEP-C concentrations with other organic carbon pools in the bathypelagic layer. Broadly, TEP-C accounts for 0.1–3% of DOC, 1–10% of high-molecular-weight DOC, and 10–>100% of the POC. Although we still have much to learn about the organic carbon stock associated with TEP, the available data suggest that TEP-C is a significant organic carbon pool in deep waters.

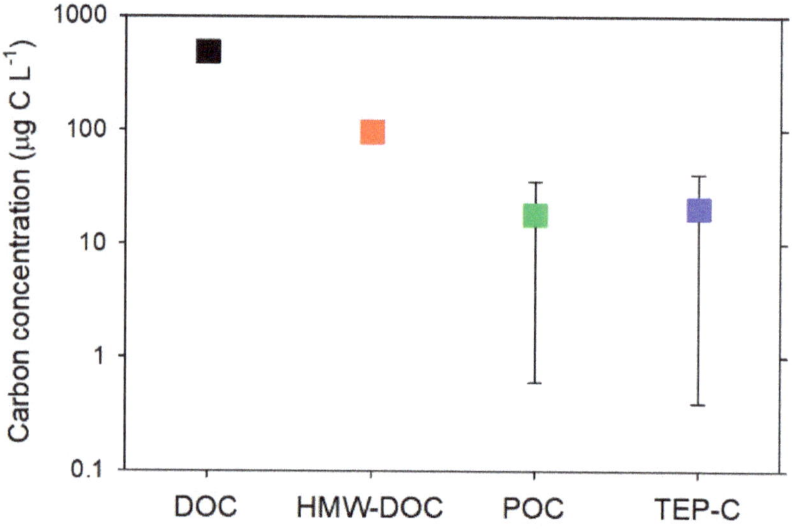

Figure 2. Concentration ranges of DOC, high-molecular-weight DOC (HMW-DOC), POC, and TEP-C in the bathypelagic layer. Error bars represent the minimum and maximum values reported in the literature and the symbols indicate the midrange values. DOC and POC values are from Nagata et al. [13]. TEP-C is from Yamada et al. [19] and Engel et al. [25]. The large ranges for POC and TEP-C reflect both regional and seasonal variabilities. In addition, there are methodological uncertainties in the POC determination (e.g., [46]) and TEP-C estimation. Although analytical errors associated with colorimetric and microscopic methods of TEP estimation are generally small (standard deviations of replicated measurements are typically <10% of the mean values [8,15,19]), there are uncertainties in the conversion factor (or equation) relating TEP to carbon (see text). The range of DOC is smaller than the size of the symbol. HMW-DOC was assumed to be 20% of DOC [34].

5. Knowledge Gaps and Future Challenges

Figure 3 summarizes the major processes involved in the TEP dynamics in the deep oceans. There are large knowledge gaps to be filled if we are to fully understand the role of TEP in the carbon cycles in meso- and bathypelagic oceans. The available data on TEP distribution at depth are scarce, especially in open oceans, hampering coherent examination of the relationship between water mass structure and TEP abundance. Data are also limited regarding the vertical and lateral transport of TEP, which severely limits our understanding of TEP dynamics in the oceans. Although the detailed, extensive examination of TEP in meso- and bathypelagic realms is a challenging task, the collected data should provide clues to evaluate the sources and sinks of TEP. Here, we list some research areas that deserve further investigation.

1. Theories have been proposed to explain the spontaneous assembly of gels [2–4] and the coagulation of particles [26] in seawater. Self-assembled gels have been identified in the deep oceanic water column [35]; however, a rigorous validation of TEP quantification methods are required to evaluate TEP formation via the spontaneous assembly of DOC. Coagulation theory is generally formulated to describe the coagulation rate as a product of particle number, collision rate, and stickiness, whereby the dominant mechanism by which the collision rate is controlled differs, depending on the particle size. Data on TEP size distributions in deep oceans are scarce [25], and we lack information about the abundance of TEP or TEP precursors in the sub-micrometer size range. Previous work has revealed that submicron particles and colloids are present in meso- and bathypelagic oceans [47–49], yet it remains to be seen if TEP are produced via the coagulation of submicron particles under deep water physical conditions. Disaggregation, the converse process of aggregation, may also affect TEP distribution at certain depths. Further studies are required to evaluate the extent of TEP delivery via the disaggregation of sinking particles.
2. To date, only a few studies have used the microscopic method to examine prokaryote colonization on TEP at depth [17,24]. These studies have provided valuable information regarding the potential role of TEP in the food webs of deep waters. Given that deep water microbial communities are dominated by organisms with surface-associated lifestyles, as evidenced by the presence of genes encoding pilus, polysaccharide, and antibiotics synthesis [36], it is likely that TEP in deep waters represent a hot spot of microbes, including prokaryotes, protists, and viruses [13]. They can also serve as important food resources for metazoan grazers that thrive throughout the oceanic water columns [50]. Despite the extensive data collected over the past two decades concerning prokaryote, protist, and virus distributions in deep water columns [12,13], further research is needed to incorporate TEP and other gel-like particles into the food web models of deep oceans.
3. To incorporate TEP dynamics into ocean carbon cycle models, it is necessary to collect quantitative data on TEP in terms of carbon. In this regard, further testing and refinement of methodologies are required to reduce large uncertainties associated with the estimation of TEP-C. It is also important to clarify the mechanisms by which TEP dynamics are regulated and to evaluate the turnover time of TEP. Currently, TEP turnover time and their lability in deep waters is poorly understood, suggesting a need to develop new methods to tackle this issue. Efforts to determine the dynamics (production and decay) of detrital polysaccharides in marine waters are inherently complicated by numerous analytical challenges [51]. We clearly need to know more about the chemical compositions, physical structures, and microbial processing of TEP and other gels in deep waters.

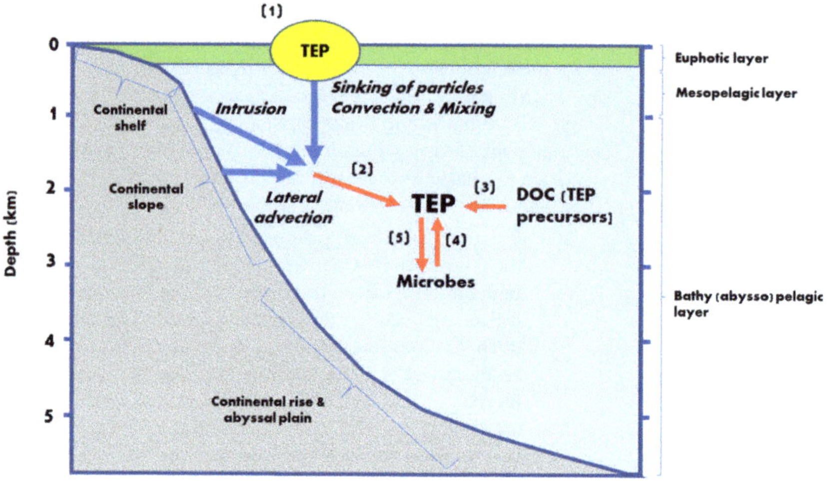

Figure 3. Schematic representation of the major processes involved in the TEP dynamics in the deep ocean. (1) TEP are mainly produced by phytoplankton and bacteria in the upper ocean [5,6]. (2) A fraction of TEP produced in the upper ocean is transported to the deeper layers. The transport processes include the sinking of particles and the advective transport due to convection and mixing, and intrusion and lateral advection. TEP transport mediated by these processes is more important in the ocean's margins than in open ocean domains. (3) TEP may be produced by the coagulation of nanogels, which are formed via the spontaneous assembly of DOC [2–4]. The ultimate source of DOC is primary production in the upper ocean, yet chemical characteristics of DOC and mechanisms underlying the persistence of DOC over millennia are poorly understood [1,38]. (4) Microbes may produce TEP, (5) while they also consume TEP. TEP are colonized by prokaryotes, acting as "hot spots" of microbes in deep oceans.

To conclude, major future challenges include the improvement of our understanding of the ocean carbon fluxes mediated by TEP at depth. The applications of gel theory and coagulation models under deep water settings are probably most effective when they are assisted by an enhanced understanding of the quantities, transport, turnover, and chemical properties of TEP. Furthermore, better recognition and determination of microbial processing of TEP are necessary for a complete understanding of TEP-mediated biogeochemical and ecological processes in deep oceans.

Author Contributions: T.N., Y.Y., and H.F. contributed to the writing of this manuscript. All authors have read and agreed to the published version of the manuscript.

Funding: This research was funded by JSPS KAKENHI Grant Numbers JP19H05667, JP17H06294, JP21H03586, JP20K19960, and JST FOREST Program Grant Number JPMJFR2070.

Data Availability Statement: Not applicable.

Acknowledgments: We thank Pedro Verdugo for inviting us to contribute to this Special Issue.

Conflicts of Interest: The authors declare no conflict of interest.

References

1. Carlson, C.A.; Hansell, D.A. DOM Sources, Sinks, Reactivity, and Budgets. In *Biogeochemistry of Marine Dissolved Organic Matter*; Hansell, D.A., Carlson, C.A., Eds.; Academic Press: Boston, MA, USA, 2015; pp. 65–126.
2. Verdugo, P. Marine Microgels. *Annu. Rev. Mar. Sci.* **2012**, *4*, 375–400. [CrossRef]
3. Chin, W.-C.; Orellana, M.V.; Verdugo, P. Spontaneous Assembly of Marine Dissolved Organic Matter into Polymer Gels. *Nature* **1998**, *391*, 568–572. [CrossRef]

4. Verdugo, P.; Alldredge, A.L.; Azam, F.; Kirchman, D.L.; Passow, U.; Santschi, P.H. The Oceanic Gel Phase: A Bridge in the Dom–Pom Continuum. *Mar. Chem.* **2004**, *92*, 67–85. [CrossRef]
5. Mari, X.; Passow, U.; Migon, C.; Burd, A.B.; Legendre, L. Transparent Exopolymer Particles: Effects on Carbon Cycling in the Ocean. *Prog. Oceanogr.* **2017**, *151*, 13–37. [CrossRef]
6. Passow, U. Transparent Exopolymer Particles (TEP) in Aquatic Environments. *Prog. Oceanogr.* **2002**, *55*, 287–333. [CrossRef]
7. Passow, U.; Carlson, C.A. The Biological Pump in a High CO_2 World. *Mar. Ecol. Prog. Ser.* **2012**, *470*, 249–271. [CrossRef]
8. Passow, U.; Alldredge, A.L. Distribution, Size and Bacerial Colonizaion of Transparent Exopolymer Particles in the Ocean. *Mar. Ecol. Prog. Ser.* **1994**, *113*, 185–198. [CrossRef]
9. Zäncker, B.; Engel, A.; Cunliffe, M. Bacterial Communities Associated with Individual Transparent Exopolymer Particles (TEP). *J. Plankton Res.* **2019**, *41*, 561–565. [CrossRef]
10. Ling, S.C.; Alldredge, A.L. Does the Marine Copepod Calanus Pacificus Consume Transparent Exopolymer Particles (TEP)? *J. Plankton Res.* **2003**, *25*, 507–515. [CrossRef]
11. Arístegui, J.; Gasol, J.M.; Duarte, C.M.; Herndld, G.J. Microbial Oceanography of the Dark Ocean's Pelagic Realm. *Limnol. Oceanogr.* **2009**, *54*, 1501–1529. [CrossRef]
12. Herndl, G.J.; Reinthaler, T. Microbial Control of the Dark End of the Biological Pump. *Nat. Geosci.* **2013**, *6*, 718–724. [CrossRef] [PubMed]
13. Nagata, T.; Tamburini, C.; Arístegui, J.; Baltar, F.; Bochdansky, A.B.; Fonda-Umani, S.; Fukuda, H.; Gogou, A.; Hansell, D.A.; Hansman, R.L.; et al. Emerging Concepts on Microbial Processes in the Bathypelagic Ocean—Ecology, Biogeochemistry, and Genomics. *Deep. Sea Res. Part II Top. Stud. Oceanogr.* **2010**, *57*, 1519–1536. [CrossRef]
14. Robinson, C.; Steinberg, D.K.; Anderson, T.R.; Arístegui, J.; Carlson, C.A.; Frost, J.R.; Ghiglione, J.-F.; Hernández-León, S.; Jackson, G.A.; Koppelmann, R.; et al. Mesopelagic Zone Ecology and Biogeochemistry—A Synthesis. *Deep. Sea Res. Part II Top. Stud. Oceanogr.* **2010**, *57*, 1504–1518. [CrossRef]
15. Passow, U.; Alldredge, A.L. A Dye-Binding Assay for the Spectrophotometric Measurement of Transparent Exopolymer Particles (TEP). *Limnol. Oceanogr.* **1995**, *40*, 1326–1335. [CrossRef]
16. Engel, A.; Passow, U. Carbon and Nitrogen Content of Transparent Exopolymer Particles (TEP) in Relation to Their Alcian Blue Adsorption. *Mar. Ecol. Prog. Ser.* **2001**, *219*, 1–10. [CrossRef]
17. Bar-Zeev, E.; Berman, T.; Rahav, E.; Dishon, G.; Herut, B.; Berman-Frank, I. Transparent Exopolymer Particle (TEP) Dynamics in the Eastern Mediterranean Sea. *Mar. Ecol. Prog. Ser.* **2011**, *431*, 107–118. [CrossRef]
18. Ortega-Retuerta, E.; Mazuecos, I.P.; Reche, I.; Gasol, J.M.; Álvarez-Salgado, X.A.; Álvarez, M.; Montero, M.F.; Arístegui, J. Transparent Exopolymer Particle (TEP) Distribution and in Situ Prokaryotic Generation across the Deep Mediterranean Sea and Nearby North East Atlantic Ocean. *Prog. Oceanogr.* **2019**, *173*, 180–191. [CrossRef]
19. Yamada, Y.; Yokokawa, T.; Uchimiya, M.; Nishino, S.; Fukuda, H.; Ogawa, H.; Nagata, T. Transparent Exopolymer Particles (TEP) in the Deep Ocean: Full-Depth Distribution Patterns and Contribution to the Organic Carbon Pool. *Mar. Ecol. Prog. Ser.* **2017**, *583*, 81–93. [CrossRef]
20. Yamada, Y.; Fukuda, H.; Uchimiya, M.; Motegi, C.; Nishino, S.; Kikuchi, T.; Nagata, T. Localized Accumulation and a Shelf-Basin Gradient of Particles in the Chukchi Sea and Canada Basin, Western Arctic. *J. Geophys. Res. Ocean.* **2015**, *120*, 4638–4653. [CrossRef]
21. Annane, S.; St-Amand, L.; Starr, M.; Pelletier, E.; Ferreyra, G.A. Contribution of Transparent Exopolymeric Particles (TEP) to Estuarine Particulate Organic Carbon Pool. *Mar. Ecol. Prog. Ser.* **2015**, *529*, 17–34. [CrossRef]
22. Cisternas-Novoa, C.; Lee, C.; Engel, A. Transparent Exopolymer Particles (TEP) and Coomassie Stainable Particles (CSP): Differences between Their Origin and Vertical Distributions in the Ocean. *Mar. Chem.* **2015**, *175*, 56–71. [CrossRef]
23. Gardner, W.D.; Jo Richardson, M.; Mishonov, A.V.; Biscaye, P.E. Global Comparison of Benthic Nepheloid Layers Based on 52 Years of Nephelometer and Transmissometer Measurements. *Prog. Oceanogr.* **2018**, *168*, 100–111. [CrossRef]
24. Busch, K.; Endres, S.; Iversen, M.H.; Michels, J.; Nöthig, E.-M.; Engel, A. Bacterial Colonization and Vertical Distribution of Marine Gel Particles (TEP and CSP) in the Arctic Fram Strait. *Front. Mar. Sci.* **2017**, *4*. [CrossRef]
25. Engel, A.; Endres, S.; Galgani, L.; Schartau, M. Marvelous Marine Microgels: On the Distribution and Impact of Gel-Like Particles in the Oceanic Water-Column. *Front. Mar. Sci.* **2020**, *7*. [CrossRef]
26. Burd, A.B.; Jackson, G.A. Particle Aggregation. *Ann. Rev. Mar. Sci.* **2009**, *1*, 65–90. [CrossRef] [PubMed]
27. Weinbauer, M.G.; Liu, J.; Motegi, C.; Maier, C.; Pedrotti, M.L.; Dai, M.; Gattuso, J.P. Seasonal Variability of Microbial Respiration and Bacterial and Archaeal Community Composition in the Upper Twilight Zone. *Aquat. Microb. Ecol.* **2013**, *71*, 99–115. [CrossRef]
28. Azam, F.; Long, R.A. Sea Snow Microcosms. *Nature* **2001**, *414*, 497–498. [CrossRef]
29. Dilling, L.; Alldredge, A.L. Fragmentation of Marine Snow by Swimming Macrozooplankton: A New Process Impacting Carbon Cycling in the Sea. *Deep. Sea Res. Part I Oceanogr. Res. Pap.* **2000**, *47*, 1227–1245. [CrossRef]
30. Martin, J.H.; Knauer, G.A.; Karl, D.M.; Broenkow, W.W. Vertex—Carbon Cycling in the Northeast Pacific. *Deep. Sea Res.* **1987**, *34*, 267–285. [CrossRef]
31. Karl, D.M.; Knauer, G.A.; Martin, J.H. Downward Flux of Particulate Organic Matter in the Ocean: A Particle Decomposition Paradox. *Nature* **1988**, *332*, 438–441. [CrossRef]
32. Passow, U.; Shipe, R.F.; Murray, A.; Pak, D.K.; Brzezinski, M.A.; Alldredge, A.L. The Origin of Transparent Exopolymer Particles (TEP) and Their Role in the Sedimentation of Particulate Matter. *Cont. Shelf Res.* **2001**, *21*, 327–346. [CrossRef]

33. Boyd, P.W.; Claustre, H.; Levy, M.; Siegel, D.A.; Weber, T. Multi-Faceted Particle Pumps Drive Carbon Sequestration in the Ocean. *Nature* **2019**, *568*, 327–335. [CrossRef] [PubMed]
34. Benner, R.; Amon, R.M. The Size-Reactivity Continuum of Major Bioelements in the Ocean. *Ann. Rev. Mar. Sci.* **2015**, *7*, 185–205. [CrossRef]
35. Ding, Y.-X.; Chin, W.-C.; Verdugo, P. Development of a Fluorescence Quenching Assay to Measure the Fraction of Organic Carbon Present in Self-Assembled Gels in Seawater. *Mar. Chem.* **2007**, *106*, 456–462. [CrossRef]
36. DeLong, E.F.; Preston, C.M.; Mincer, T.; Rich, V.; Hallam, S.J.; Frigaard, N.U.; Martinez, A.; Sullivan, M.B.; Edwards, R.; Brito, B.R.; et al. Community Genomics among Stratified Microbial Assemblages in the Ocean's Interior. *Science* **2006**, *311*, 496–503. [CrossRef]
37. Vidal-Melgosa, S.; Sichert, A.; Francis, T.B.; Bartosik, D.; Niggemann, J.; Wichels, A.; Willats, W.G.T.; Fuchs, B.M.; Teeling, H.; Becher, D.; et al. Diatom Fucan Polysaccharide Precipitates Carbon During Algal Blooms. *Nat. Commun.* **2021**, *12*. [CrossRef]
38. Hansell, D.A. Recalcitrant Dissolved Organic Carbon Fractions. *Annu. Rev. Mar. Sci.* **2013**, *5*, 421–445. [CrossRef]
39. Bercovici, S.K.; Arroyo, M.C.; De Corte, D.; Yokokawa, T.; Hansell, D.A. Limited Utilization of Extracted Dissolved Organic Matter by Prokaryotic Communities from the Subtropical North Atlantic. *Limnol. Oceanogr.* **2021**. [CrossRef]
40. Shen, Y.; Benner, R. Molecular Properties Are a Primary Control on the Microbial Utilization of Dissolved Organic Matter in the Ocean. *Limnol. Oceanogr.* **2019**, *65*, 1061–1071. [CrossRef]
41. Nagata, T.; Fukuda, H.; Fukuda, R.; Koike, I. Bacterioplankton Distribution and Production in Deep Pacific Waters: Large-Scale Geographic Variations and Possible Coupling with Sinking Particle Fluxes. *Limnol. Oceanogr.* **2000**, *45*, 426–435. [CrossRef]
42. Stief, P.; Elvert, M.; Glud, R.N. Respiration by "Marine Snow" at High Hydrostatic Pressure: Insights from Continuous Oxygen Measurements in a Rotating Pressure Tank. *Limnol. Oceanogr.* **2021**. [CrossRef]
43. Tamburini, C.; Boutrif, M.; Garel, M.; Colwell, R.R.; Deming, J.W. Prokaryotic Responses to Hydrostatic Pressure in the Ocean—A Review. *Environ. Microbiol* **2013**, *15*, 1262–1274. [CrossRef] [PubMed]
44. Fukuda, R.; Ogawa, H.; Nagata, T.; Koike, I.I. Direct Determination of Carbon and Nitrogen Contents of Natural Bacterial Assemblages in Marine Environments. *Appl. Environ. Microbiol* **1998**, *64*, 3352–3358. [CrossRef]
45. Tanet, L.; Martini, S.; Casalot, L.; Tamburini, C. Reviews and Syntheses: Bacterial Bioluminescence—Ecology and Impact in the Biological Carbon Pump. *Biogeosciences* **2020**, *17*, 3757–3778. [CrossRef]
46. Turnewitsch, R.; Springer, B.M.; Kiriakoulakis, K.; Vilas, J.C.; Arístegui, J.; Wolff, G.; Peine, F.; Werk, S.; Graf, G.; Waniek, J.J. Determination of Particulate Organic Carbon (POC) in Seawater: The Relative Methodological Importance of Artificial Gains and Losses in Two Glass-Fiber-Filter-Based Techniques. *Mar. Chem.* **2007**, *105*, 208–228. [CrossRef]
47. Koike, I.; Hara, S.; Terauchi, K.; Kogure, K. Role of Sub-Micrometer Particles in the Ocean. *Nature* **1990**, *345*, 242–244. [CrossRef]
48. Nagata, T.; Koike, I. Marine Colloids: Thier Roles in Food Webs and Biogeochemical Fluxes. In *Biogeochemical Processes and Ocean Flux in the Western Pacific*; Sakai, H., Nozaki, Y., Eds.; Terra Scientific Publishing Company (TERRAPUB): Tokyo, Japan, 1995; pp. 275–292.
49. Wells, M.L.; Goldberg, E.D. The Distribution of Colloids in the North-Atlantic and Southern Oceans. *Limnol. Oceanogr.* **1994**, *39*, 286–302. [CrossRef]
50. Hernandez-Leon, S.; Koppelmann, R.; Fraile-Nuez, E.; Bode, A.; Mompean, C.; Irigoien, X.; Olivar, M.P.; Echevarria, F.; Fernandez de Puelles, M.L.; Gonzalez-Gordillo, J.I.; et al. Large Deep-Sea Zooplankton Biomass Mirrors Primary Production in the Global Ocean. *Nat. Commun.* **2020**, *11*, 6048. [CrossRef] [PubMed]
51. Arnosti, C.; Wietz, M.; Brinkhoff, T.; Hehemann, J.H.; Probandt, D.; Zeugner, L.; Amann, R. The Biogeochemistry of Marine Polysaccharides: Sources, Inventories, and Bacterial Drivers of the Carbohydrate Cycle. *Ann. Rev. Mar. Sci.* **2021**, *13*, 81–108. [CrossRef]

gels

MDPI

Review

Marine Polymer-Gels' Relevance in the Atmosphere as Aerosols and CCN

Mónica V. Orellana [1,2,*], Dennis A. Hansell [3], Patricia A. Matrai [4] and Caroline Leck [5]

1 Polar Science Center, Applied Physics Laboratory, University of Washington, Seattle, WA 98195, USA
2 Institute for Systems Biology, Seattle, WA 98109, USA
3 Department of Ocean Sciences, RSMAS, University of Miami, Miami, FL 33149, USA; dhansell@miami.edu
4 Bigelow Laboratory for Ocean Sciences, East Boothbay, ME 04544, USA; pmatrai@bigelow.org
5 Department of Meteorology, Stockholm University, 11419 Stockholm, Sweden; lina@misu.su.se
* Correspondence: morellan@uw.edu

Abstract: Marine polymer gels play a critical role in regulating ocean basin scale biogeochemical dynamics. This brief review introduces the crucial role of marine gels as a source of aerosol particles and cloud condensation nuclei (CCN) in cloud formation processes, emphasizing Arctic marine microgels. We review the gel's composition and relation to aerosols, their emergent properties, and physico-chemical processes that explain their change in size spectra, specifically in relation to aerosols and CCN. Understanding organic aerosols and CCN in this context provides clear benefits to quantifying the role of marine nanogel/microgel in microphysical processes leading to cloud formation. This review emphasizes the DOC-marine gel/aerosolized gel-cloud link, critical to developing accurate climate models.

Keywords: marine gels; DOC; aerosols; CCN; SML; central Arctic Ocean

Citation: Orellana, M.V.; Hansell, D.A.; Matrai, P.A.; Leck, C. Marine Polymer-Gels' Relevance in the Atmosphere as Aerosols and CCN. *Gels* **2021**, *7*, 185. https://doi.org/10.3390/gels7040185

Academic Editor: Jean-Michel Guenet

Received: 9 September 2021
Accepted: 23 October 2021
Published: 28 October 2021

Publisher's Note: MDPI stays neutral with regard to jurisdictional claims in published maps and institutional affiliations.

1. Introduction

Dissolved organic carbon (DOC) is the largest reservoir of reduced organic carbon in the world's oceans (~662 Pg C; [1]), similar in size to the atmospheric reservoir of CO_2 (750 Pg C [2,3]). Of autotrophic origin mostly, this enormous biopolymeric pool is consumed by heterotrophic bacteria and protozoa [4]. Applying the principles of soft matter physics to understand marine biopolymer dynamics, Chin et al. [5] demonstrated that marine biopolymers assembled into a distinct supramolecular organization forming microscopic 3-dimensional polymer-gel networks embedded in a solvent (seawater) and ranging in size from nanometers to microns [5]. Spontaneous assembly of marine polymer gels occurs in the oceans when a chemically heterogeneous, polydispersed mixture of organic biopolymers (proteins, carbohydrates, lipids, and nucleic acids) interacts to form randomly tangled 3D, cross-linked hydrated networks. These are held together by ionic bonds (Ca^{+2}), hydrophobic bonds, hydrogen bonds, and/or van der Waal forces, with a characteristic assembly/dispersion equilibrium depending on the nature of the polymers and the relation with the solvent [5–9]; for details please see Verdugo [10] in this volume.

Marine polymer assembly is reversible, follows second/first order kinetics, and exhibits an in vitro approximate thermodynamic yield at equilibrium of 10% (25–30% in the central Arctic Ocean) in surface waters and lower in deep waters (3–5%); thus most likely, ~70 Pg C of DOC [9] remains in dynamic and reversible assembly equilibrium, forming porous networks [5,7] in a colloidal size continuum [9]. As such, marine polymer gels have been proposed to play a pivotal role in regulating ocean basin-scale biogeochemical dynamics [11]. Marine gels concentrate and accumulate in the sea surface microlayer [7,12,13], and are then available for air–sea exchange as organic aerosols, as observed in the central Arctic Ocean [7,13–15] and, potentially, as a source for cloud condensation nuclei (CCN) formation [7,15]. Marine primary (aerosols emitted directly into the atmosphere) organic aerosols are airborne nanometer size to micrometer size particles (liquid, two phase liquid, solid) in the atmosphere that have been additionally quantified over the Atlantic and the Pacific Oceans and other oceans as well [16–19]. Thus, understanding marine biopolymer dynamics is critical to developing accurate models of the response of oceanic and atmospheric biogeochemical cycles to climate change. One crucial area is understanding the role of marine biopolymer nanometer size gels in cloud formation, and the link between the ocean's surface biology to the atmosphere and climate [7,19–22]. Understanding the gel's sources, their emergent properties (assembly, volume phase transitions), composition, fluxes, and size distributions (among other characteristics) is necessary to assess their susceptibility to influence cloud formation processes.

This article introduces the critical role of marine gels as a source of aerosols and CCN in cloud formation processes, emphasizing Arctic marine microgels. While there are many studies and reviews about organic aerosols and cloud formation, understanding organic aerosols and CCN in the context of soft matter physics can provide clear benefits to quantifying their role in microphysical processes leading to cloud formation. Understanding the response of biogeochemical cycles to environmental forcing, and specifically the DOC –marine gel/aerosolized gel-cloud link, is critical to developing accurate climate models (Figure 1).

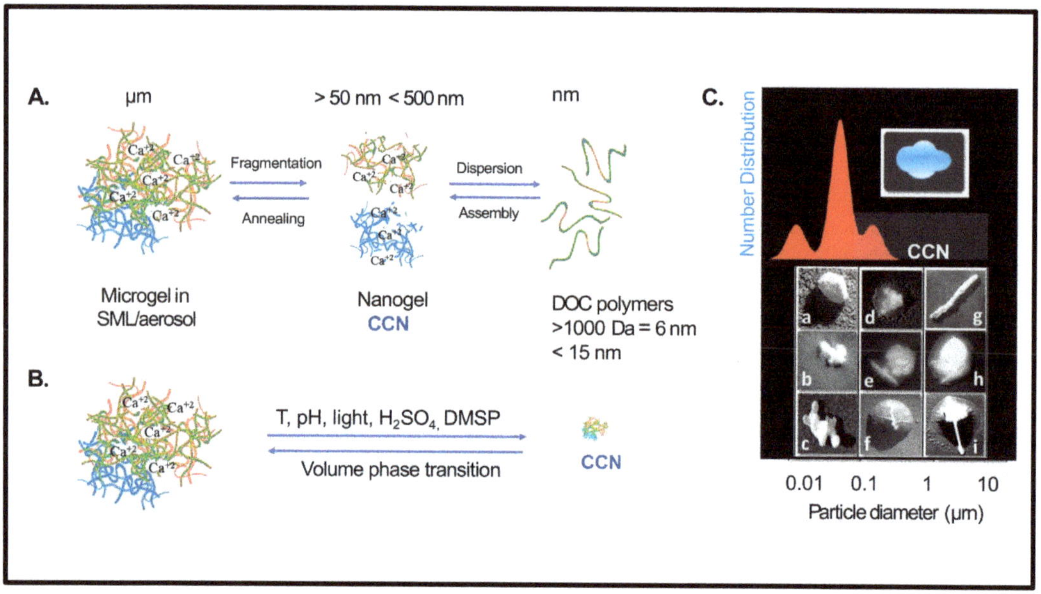

Figure 1. Conceptual figure indicating dynamic processes affecting gels as aerosols, CCN, and free polymers. (**A**) At the ocean–air interface in the surface microlayer, dissolved organic carbon (DOC) polymers assemble in a reversible process into microgels stabilized by entanglements and Ca^{+2} bonds and/or hydrophobic moieties. Microgels are then available for air–sea exchange as organic aerosols by diverse processes (bubble bursting/wind). Aerosolized microgels can fragment into smaller size nanometer-size gels by UV exposure, and/or dispersion, or other processes. If the nanometer size gels are activated, they can nucleate into forming CCN. Furthermore, nanogels can further disperse into DOC free polymers, (adapted from Verdugo, 2012). (**B**) Microgels can also attain nanometer sizes by undergoing volume phase transition induced by environmental conditions such as changes temperature (T), pH, light, H_2SO_4, DMSP, and DMS. These nanometer size gels may also become CCN, however, this route has not been proven yet. (**C**) TEM pictures of aerosol particles collected over the central Arctic Ocean north of 80°N. Examples of the changing nature of the high Arctic particles in different modal diameters: (**a–c**) sub-Aitken mode, (**a**) penta-hexagonal structure, crystalline and hydrophobic in nature assumed to be

a colloidal building block of a polymer gel, (**b**) small polymer gel-aggregate forming a "pearl necklace" morphology possibly indicating hydrophobicity, according to Saiani et al. (2009), slightly covered with hydrophilic viscous but not gelling polymeric material "mucus," (**c**) another particle example similar to b, (**d–f**) Aitken to small accumulation mode, (**d**) particle with a high sulfuric acid content and with a gel-aggregate inclusion embedded in a viscous non-gelling film of high organic content, (**e**) gel-aggregate, and a particle resembling a bacterium with a small aggregate attached to it, possibly detached from the larger one. The "bubble-like shaped particles" may indicate a possible recent injection to the atmosphere at the air–sea interface, (**f**) particle containing mainly ammonium sulfate and methane sulfonate, (**g–i**) large accumulation mode, (**g**) a bacterium, (**h**) sea-salt with an organic content only present at the rare occasion of high winds > 12 m s^{-1}, (**h**) sea salt and a bacterium coated with an organic film and by the concentric rings typical of droplets of sulfuric acid.

2. Background: Marine Gel Relevance as Aerosols

Our knowledge about clouds endures as a limiting factor in our understanding of the climate system and consequently in climate modeling [23,24]. Clouds only form when water vapor condenses. However, in the atmosphere water vapor needs a substrate onto which to condense on—tiny airborne aerosol particles known as CCN. Typically, CCN fall within the submicron size fraction, ~100 nanometers in spherical diameter. Depending on their properties and heights, clouds can either warm the underlying Earth surfaces by triggering a localized greenhouse effect or cool them by outwardly reflecting solar radiation. If CCN are limited and sparse, the resulting clouds will contain fewer and larger droplets [25]. Such clouds will reflect and scatter less sunlight radiation into space while blocking the escape of heat from Earth's surface, causing it to warm [26].

When CCN are plentiful [27], countless fine droplets form; the resultant clouds will scatter additional light and become better reflectors, thus cooling the surface below. Anthropogenic particles are essentially absent in the central Arctic (>80° N), especially during summertime. Instead, biological sources of particles may dominate [7,15,21,28–30]. This "clean" air, with few CCN, makes the low-level stratocumulus clouds optically thin, with fewer but larger droplets. Because of the direct link between production of organic carbon—specifically marine gels by microorganisms—and CCN [7,15], climate change, ocean warming, and acidification may affect the microbiota's diversity and activity, directly affecting the production of gels and, hence, aerosols and cloud formation [31]. Over the last decades or so, research extending the high Arctic findings to lower latitude oceans, has stressed the presence and enrichment of marine organic matter particles of submicron sizes in airborne aerosols and cloud water [13,15,18,20,22,28,32–45]. Although these organic aerosol particles are commonly described in terms of the chemical composition of the size-segregated organic components of the marine primary aerosols or the functional group composition (i.e., low molecular weight carboxylic acids, other humic-like substances) (e.g., [18,46]), they most likely are hydrated polymeric species and most likely form hydrated nanonetworks, or marine nanosized hydrogels, as they derive directly from marine dissolved organic matter.

Exopolymer-like particles in the atmosphere were first discovered by Bigg and Leck [15,28,45,47]. These authors recognized that these particles displayed the physicochemical characteristics of "marine gels", polymer networks with emergent properties (please see below), which was confirmed by Orellana et al. [7]. This understanding followed from their studies of a possible link between cloud formation and biopolymer gels, then described as exopolymer substances (EPS) in the surface microlayer (SML) (<100 µm thick at the air–sea interface) in the high Arctic sea-ice leads [14,31,48].

Exopolymer-containing marine particles are also known as transparent exopolymer particles (TEP) when operationally defined as particles stained with Alcian blue, a cationic copper-phthalocyanine dye dissolved in acetic acid at pH 2.5 [49,50] that preferentially stains COO^- acidic and half-ester sulphate reactive groups of acidic polysaccharides [51] and uronic acids [52,53]; or stained with Coomassie blue (CB) for particles containing proteins. TEP and CB particles have also been measured in the SML [54] and atmosphere in the North Atlantic [55]. However, the fixation with acetic acid (TEP) or citric acid at low pH (CB) changes the macromolecular conformation and physical dynamics of the particles.

Charlson et al. (1987) evaluated the existing evidence at the time linking the gas dimethyl sulfide (DMS) (produced by microbial food web interactions from its precursor dimethylsulfoniopropionate (DMSP) to the production of CCN over remote marine areas. This challenging "CLAW hypothesis" [56] proposed that, in the surface ocean, DMS gas emissions by phytoplankton and their subsequent known oxidation products in the atmosphere—methane sulfonic acid, sulfur dioxide, and sulfuric acid—trigger cloud formation, in turn cooling the ocean surface. This cooling effect would, in turn, affect further emissions of DMS by changing the speciation and abundance, or both, of marine phytoplankton, thus establishing a negative or stabilizing feedback loop. High Arctic observations in the early 1990s confirmed that the intermediate oxidation products provided most of the mass for the CCN-sized particles observed over pack ice [57]. The source location of most of the DMS, though, was found at the fringe of the central Arctic Ocean, at the hospitable edges of the pack ice (or marginal ice zone), not in the central area of the Arctic Ocean [57]. The distribution suggested that winds carried DMS-rich air at the edges of the pack ice [29] towards the North Pole, and oxidation of the airborne DMS created extremely small sulfuric acid-containing particles. Theoretically, these particles would grow slowly by further condensation of the acids until they were large enough to serve as CCN. Surprisingly, sulfuric acid was not involved with the production of the small precursors of CCN [58], which was later also confirmed by Quinn and Bates [41]. Instead, the high Arctic observations in the mid-1990s showed that these small precursors were organic material and mostly particles resembling bacteria and nanometer- and micron- sized gels that were accompanied by other larger particles, such as fragments of diatoms [15,28,37,45,59]. Subsequently, Bigg et al. [48] detected large numbers (10^6–10^{14} mL^{-1}) of similar particles within the SML between ice floes.

During the summer in 2008, at 87° N [60], when the ice melt in the central Arctic Ocean was maximal, and the Arctic ice leads were most prevalent, marine polymer gel's identification, characterization, and quantification were conducted in seawater, and, for the first time, also in the SML, fog, cloud, and aerosols [7]. This work confirmed that assembled nano-, micro-, and fewer macro-sized gels found in the SML were similar to earlier findings [15]. Orellana et al. [7] also found that the airborne microgels may have the chemical surfactant properties necessary to act as CCN. However, to behave as effective CCN, these particles must first reach a critical size (equal or larger than 50 nm [20]) and meet other physicochemical properties and energy constraints. Leck and Bigg [37,59] and Karl et al. [61] speculated that the primary marine gel would disintegrate under some circumstances, generating smaller sized particles, most likely due to ultraviolet (UV) radiation cleavage [62], or they might disperse under the different physicochemical conditions (temperature, pH, ionic strength) present in the atmosphere than in the seawater. However, polymer gels may also attain smaller sizes by undergoing volume phase transition [5], also quantified in the Arctic surface waters [7]. Furthermore, it was demonstrated that microgels could also carry DMSP similarly to phytoplankton secretory vesicles [63], or be the site for DMS condensation [36]. Orellana et al. [7] additionally demonstrated the presence of peptide amphiphiles as important characteristic components of such polymer nano- and micro-gels. However, Martin et al. [64] suggested that amphiphilic biopolymeric gels cannot uptake water vapor and form cloud droplets and activate as CCN, most likely due to the effects of surface partitioning on the lowering surface tension not being taken into account in their calculations [65]; instead, they concluded the sulfate fraction of the particles that dominated the CCNs, possibly trapped within the polyanionic matrix of the gel [63]. Similarly, Ovadnevaite et al. [16] suggested a dichotomous behavior for the primary mixed surface marine organic aerosols, most likely due to the lowering of the surface tension [66], and inhibition of water uptake by the hydrophobic surface of the organic-rich gel particles, as demonstrated experimentally [67]. Other measurements in the high Arctic have also suggested organic aerosols-like gels, as well as sea salt aerosols and older, long-range transported continental aerosols [30,68,69]. Recently, Baccarini et al. [70] demonstrated that frequent new particle formation, which could potentially lead to CCN over the high Arctic pack ice, is enhanced by iodine emissions, most likely produced by the microbiota [71]. Iodination of natural organic matter involves iodination of aromatic moieties of humic substances (HS) and proteins that could be aerosolized [72]. Organo iodine is very abundant in the ocean [73] and, in the Arctic Ocean, HS are more abundant in the winter than in the spring-summer months [74]. Perhaps, both iodine emissions and nanogels impact the microphysical properties of the clouds over the central Arctic Ocean [75]. In fact, while some fraction of the hypoiodous acid (HOI) production by marine diatoms can be volatile, another fraction may also react with seawater DOC polymers, thus constituting a critical mechanism to transfer this chemical to the atmosphere via aerosolized gels [71].

3. Composition and Controls on Microgel Formation and Bioreactivity

DOC is biopolymeric [4] and phytoplankton and bacteria are the main producers [76–78]. DOC is operationally categorized into three major fractions according to its apparent biological lability [79]. All ocean depths and geographical areas contain the very old, biologically refractory DOC (RDOC concentrations < 45 μM, with bulk radiocarbon ages of >6000 years, [80]); a semi-labile fraction that accumulates in the surface ocean and mixes towards the ocean interior (10–30 μM), and a labile fraction produced daily at the ocean surface by autotrophs and degraded swiftly by heterotrophs (hours, days, months). Gel forming EPS (colloidal and macromolecular size) in the Gulf of Mexico and the Atlantic Ocean have been shown to have modern radiocarbon ages [81].

Phytoplankton produces DOC [7,12,82] by diverse mechanisms [83], including direct release [84,85], mortality by viral lysis [86,87], apoptosis [88–90], degradation of particulate organic matter by microbes [91], and grazing [92,93]. Phytoplankton alone release ~10–30% of their primary production into the DOC pool by regulated exocytosis and/or cell death [90] in the form of microgels and free biopolymers [5,94], also known as secretions, organic surfactants, and exopolymers. These contain carbohydrates [95–97], peptides and proteins [98], lipids [99,100], and other metabolites [82,101,102]. The DOC pool, in turn, drives heterotrophic bacterial growth and marine ecosystem dynamics [103–105]. Furthermore, ~50% of bacterial production is also released into the DOC pool by viral burst and mortality [87], most likely releasing bacterial membrane porins [106,107], hydrolases [108], fatty acids and lipopolysaccharides [100]. Bacteria also release refractory short-chain compounds [109,110].

In the central Arctic Ocean, polymer gels are produced from the biological secretions of marine phytoplankton [111–115], bacteria and sea ice algae (reviewed by Deming and Young (2017) [116] and references there in), as well as from cell debris [117]. These polymers accumulate at the SML, the upper most layer of the ocean (10–1000 mm thick [118]), where they are available for aerosolization and, eventually, cloud formation [7,37]. These polymers are rich in polysaccharides [119] and macromolecules such as ice- binding proteins, nucleic acids, lipids, phenols, and flavones [120]. Bubble bursting in the SML transfers polymer-gels/aerosols into the atmosphere [18,118,121–123]. The SML thus has a crucial role in several biogeochemical cycles, such as the carbon cycle, the transfer of gases and aerosols to the atmosphere and climate-related processes [7,14,19,31,54,118].

Phytoplankton community composition exhibits seasonal changes influencing the DOC composition and abundance [119,124], and thus affecting the aerosol composition, reactivity, and particle growth [39]. In winter, the aerosols composition is dominated by sea salt (83%) and other inorganic compounds (non-sea salt $SO4^{2-}$, metasulfonic acid (MSA), and low organic matter (5–15%) [39,125]. In contrast, organic matter dominates in aerosols during the spring bloom (40–65%) with a low percentage of sea salt and other inorganic compounds [39,125], as well as during the subsequent bloom collapse by viral infection [22,126,127]. Likewise, the chemical composition of simulated microgels strongly influences the activation capacity and growth of aerosolized microgels to act as CCN [20], linking the gel's marine source to their atmospheric role.

The aerosol's chemical composition influences the microscale physical processes in the atmosphere, and their capacity to produce CCN. Therefore, the aerosol's composition has been characterized by different methods ranging from real-time aerosol mass spectrometers, aerosol time of flight mass spectrometer (ATOFMS), Fourier-transform infrared spectroscopy, single particle spectroscopy, as well as chemical fractionation and analysis among others. Frossard et al. [128,129] reviewed several studies and methods used to determine organic matter composition and particle size from natural and bubbler-generated, nascent sea-spray aerosols. However, it is not clear that the same population and type of organic particles were sampled by these different studies [130,131]. Furthermore, these studies took place at different places and different times of the year when the phytoplankton populations producing organic material were different.

Chemical composition strongly influences the hygroscopicity activation capacity and growth of primary organic aerosols to act as CCN, determined experimentally as hydrophobicity, volatility, changes in surface tension, surface charge, or solubility ([46] and references therein). This is critical, as the capacity of marine DOC polymers in the surface ocean to assemble into gel particles (see below) is also established by ionic bonds, polymer persistence length, and charge density, as mentioned earlier [5,9,11,62]. However, the chemical composition of dispersed aerosol components may be very different from that of assembled gel particles [7], which also exhibit emergent properties (i.e., reversible volume phase transition from a swollen, hydrated phase to a condensed and compact phase and vice-versa, see below) that arise from their interactions with seawater, or their control by environmental stimuli in seawater. These gel properties could also determine their characteristics as CCN.

Size-specific measurements show that aerosol composition as well as CCN activity vary with aerosol size [20,39,41] could dispersion of the gels during analysis produce this finding? However, over the Pacific and the Atlantic Oceans, marine organic aerosols dominate the nanometer sized aerosols composition [16,17,66,132]. Because mass spectrometers can only analyze <1 μm size particles, it is thought that nanometer sized dissolved carbon that is refractory and of old age dominates the composition of sea spray and primary aerosols [17,18,133]. Conversely, nanometer sized aerosols could be the result of cleaved, dispersed organic and labile micron sized DOC gels, and perhaps mixed with old refractory nanometer sized gels, and hence not completely old and refractory (19–40%), as shown by radiocarbon aging and by the thermal stability of the organic material [7,133,134]. Thus, this question remains to be revisited.

4. DOC and Gels: Assembly of Biopolymers

Chin et al. [5] applied the principles of soft matter physics to understand marine biopolymer dynamics, demonstrating that marine biopolymers assemble into 3D gel networks. As indicated earlier, spontaneous assembly of marine polymer gels occurs in the oceans when a poly-dispersed mixture of marine biopolymers interacts to randomly form tangled 3D cross-linked networks, held together by ionic bonds (Ca^{+2}), and/or hydrophobic forces, hydrogen bonds, Van der Waals forces, depending on the nature of the polymers and the relation with the solvent (in this case seawater) [5–9]. Marine polymer assembly is reversible, follows second or first order kinetics depending on the composition of the polymers [5–7], and exhibits an approximate thermodynamic yield at equilibrium of 10% in open ocean waters. During this assembly process, biopolymers tangle and anneal to form nanometer to micrometer sized, porous networks in 48 h in laboratory conditions which remain in dynamic equilibrium [5,7,9], as part of a colloidal size continuum [9]. The yields of gel assembly depend on the polymer length (see below), composition (see above), charge density, and the presence of hydrophobic moieties [6,98]. In high Arctic surface ocean waters, DOC polymer assembly follows first order kinetics, with an approximate thermodynamics yield at equilibrium of 25–30%. The difference in the kinetics of polymer assembly between geographical regions arises from the composition of the DOC biopolymers. In the Arctic Ocean, the biopolymers are amphiphilic, with hydrophilic biopolymers containing hydrophobic moieties [6,98].

Assembled microgels accumulated in the SML are then available for air sea exchange as organic aerosols and potentially as a source of CCN [7] in the Arctic Ocean as well as in other oceans [18,22,42]. During cloud formation, aerosols containing polymers uptake large amounts of water, hydrating and swelling, most likely facilitating the production of microdroplets. However, the assembly of aerosolized and hydrated gels is yet to be quantified and it could likely explain not only the formation of microdroplets but also the production of secondary marine organic gel particles in aerosols [37].

5. Microgel Size and Stability: Dependency on Polymer Length

Polymer theory states that the probability of assembly of polymers into polymer gels, their equilibrium size and their stability as tangled networks once assembled, increases with the square of the polymer length [135,136]. DOC polymers of greater length are able to assemble into stable and larger size polymer gel networks because the interactions between the polymer chains tangles (i.e., ionic, van der Waals, hydrophobic forces, etc.) and interpenetrations become stronger. However, nanometer sized (<1000 DA) molecules are either unable to assemble or they assemble into unstable and colloidal size networks due to the low degree of interaction between the polymer chains; thus, polymer length constitutes a central control on the primary organic aerosol ultimate size spectrum and on the necessary characteristics for being CCN (≥50 nm). This theory has multiple implications for cloud dynamics, and offers insights on the sources and sinks of cloud particles [11,62,135–137].

As stated above, it has been demonstrated that the assembly of polymer gels in 0.2 μm-filtered seawater exposed to UV-B radiation levels found in polar regions [138], or to bacterial degradation, was slower than in unfiltered controls [62,139]. These results suggest that photochemical or bacterial enzymatic degradation of polymers can drastically limit the supply of microgels of bigger sizes (>500 nm); UV-B-induced cleavage (in vitro and in the field) and biodegradation yield short-chain polymers that do not assemble or assemble into unstable colloidal (nanometer size) gels. However, proteins exposed to photooxidation and added reactive oxygen species at levels measured at the ocean surface and room temperature (22 °C) do not cleave into short polymers chains but instead aggregate, preserving the original proteins [140]. However, in Arctic waters at 87° N, microgels did cleave when irradiated with environmental levels of UV radiation and cold temperature (−2 °C, −4 °C [60,141]). Indeed, exposure to environmental levels of UV radiation resulted in a factor of three reduction of the marine microgel yield, indicating cleavage and dispersion of the gels. While reactive oxygen species where not measured in the Arctic Ocean experiments, the difference in ambient temperature (22 °C and −4 °C) may explain the different processes taking place in both of these measurements [7,140]. In fact, polar phytoplankton and bacterial species develop antifreeze proteins with distinct structural and antioxidant properties, as well as amino-acid composition [142], than the proteins (i.e., fetuin, bovine serum albumen, cytochrome c) tested by Sun et al. [140].

Polymer gels range from nanosized networks to a few microns [7,15] in the central Arctic Ocean. Small nanometer sized aerosol gels, which are the most abundant, can result from degradation and dispersion during long travel times over open waters, or over the pack ice that, when combined with freshly-produced long aerosolized DOC polymers, facilitate the simultaneous production of multi-sized aerosol particles [37,59,61], rather than the traditionally-observed chemical and/or condensational progressive particle growth in the atmosphere. UV-B radiation can also disperse already assembled gels when crosslinked polymers are degraded and cleaved by UV-B [62].

6. Volume Phase Transition: Effects of pH, DMS, and DMSP on Gel Dynamics

The assembly and dispersion of assembled macromolecules are also affected by environmental forcing, such as temperature [9,143], pressure, pH [5,7], DMSP and DMS concentration [7], polycations such as polyamines [144], trace metals [145], pollutants [146], electric fields [147], and light [148]. Changes in the previous parameters stimulate polymer gel volume phase transitions [149] (swelling or dehydration and condensation of the polymer gel networks) and further the collapse of microgels into a dense polymeric nanometer network. In the ocean, volume phase transition of polymers gels may increase the sedimentation rate of the gels into deeper [5,7,12,150], thus removing them as a source of aerosols at the surface. During gel dehydration and collapse, small molecules can be entrapped in the gel network at high concentrations, such as DMSP [63] or even proteins (RubisCO) that can be found unaltered in the deep ocean [150]. Volume swelling/dehydration of polymer gels may be an important process in the central Arctic Ocean, where nanomolar concentrations of DMSP and DMS as well the products of DMS oxidation, such as sulfuric acid, induce phase transition of the polymer gels in vitro [7].

New data indicating the presence of HIO having a role in new particle, and possibly CCN, formation [70] should be tested for a predictable phase transition of micron sized gels. Arctic Ocean microgels could also entrap iodine and its products [71], as they do for DMS [63]. We also predict Arctic Ocean aerosolized polymer gels to be sensitive to changes in temperature [9], where supercool temperatures in the clouds [141] could induce volume phase transition of polymer gel networks. Changes in temperature can also change the hygroscopic properties of the polymer gel network due to the induced phase changes of the polymer network within the spray drop, and thus facilitate the formation of rain drops. However, this must be explored and demonstrated.

Furthermore, the presence of trace metals in the central Arctic Ocean [151–153] could also induce gel volume phase transition.

7. Summary of Present Knowledge: Marine Polymer Gels as CCN in the Central Arctic Ocean

During the Arctic Summer Cloud Ocean Study (ASCOS 2008) study at 87° N [60], when melt was maximal and leads were most prevalent, identification, characterization and quantification of marine polymer gels in seawater and also in the surface SML, fog, cloud, and aerosols were performed [7]. These polymer gel networks reached high concentrations in seawater (10^6–10^9 mL^{-1}), with assembly yields averaging 25–30%, which were higher than published previously [5,62,154]. Their sizes ranged from a few microns to nanometer size, with nanometer size polymer networks being the most abundant [7]. A specific fluorescently labeled antibody probe developed against in situ seawater and SML biopolymers confirmed for the first time that the particles found in the atmosphere (aerosol/fog/cloud) originated in the surface sea water, including the subsurface and SML [7]. The polymeric particles were likely released by the abundant sea-ice diatoms (*Melosira arctica* and *Fragilariopsis cylindrus*), other phytoplankton and bacteria, and behaved as nano- and micron size gels, demonstrating a direct link between CCN and microorganisms and, more specifically, marine biogenic polymers or marine gels [7,15]. The gel networks were held together by random entanglements and Ca^{+2} ionic bonds as well as by hydrophobic moieties. The gels comprised as much as 50% of the total dissolved organic carbon in surface waters and the SLM, and they assembled at 4 °C faster than previously observed, following first order kinetics, most likely due to the presence of hydrophobic moieties enhancing polymer assembly [6,155]. The marine gels also underwent volume phase transitions induced by DMSP as well as DMS, another indication that those particles displayed the physicochemical characteristics of gels. Gel abundance in seawater also correlated with enrichment of proteins containing hydrophobic amino acids (leucine, isoleucine, phenylalanine, and cysteine) and of DMSP in the SML [7]. The aggregates found in the SML were confirmed to be similar to earlier observations [15,48].

The surface activity of aerosol particles, specifically the effect on surface tension reduction and its effect on the equilibrium spherical radius of an aqueous drop, can significantly influence the cloud droplet forming ability of these particles. In an attempt to reduce some of the uncertainties surrounding the observed CCN properties promoting/suppressing cloud droplet formation over the Arctic pack, Leck and Svensson [20] used Köhler theory and simulated the cloud nucleation process using an adiabatic air parcel model that solves the kinetic formulation of water condensation on aerosol. They took advantage of highly size resolved impactor samples of inorganic water-soluble aerosol-bulk chemistry together with size-resolved electron microscope aerosol particle data collected on previous expeditions [7,15,36,45,156,157]. This simulation made possible a highly size resolved best "guess" of the unexplained particle number fraction assumed to consist of organic water soluble, slightly water soluble and non-water-soluble proxy constituents. The general conclusion from the simulations was the increase of a hydrophobic character, with decreasing diameter, of the activated particles. This suggested a hydrophobic character for the central Arctic Aitken-mode (15–80 nm diameter) aerosols that would in turn impede water uptake and suppress cloud activation below 0.4% water supersaturation. As such, it seems that very high water supersaturations (>0.8%) would be required in order for the Aitken mode particles to be activated. It is possible that such high water supersaturations occur where small total droplet number concentrations are present such that excess water vapor is not depleted by larger particles and helps sustain the cloud even when the Aitken particles have low hygroscopicity [158].

The results from the above studies are consistent with the dichotomous behavior [16] of the 3D structure of the polymer gels during cloud droplet activation. Initially, only partial wetting and only weak hygroscopic growth would occur since only part of the CCN surface exhibits strong hydrophilicity. Given time, the strong surfactant property of the gel hydrophilic entities would decrease surface tension, which would lead to a decrease in water vapor supersaturation necessary to promote cloud droplet formation [137].

Not only could the polymer gel surface facilitate nucleate cloud droplets, but it is possible that the protein amino acid sequences [159] may play an active role, i.e., through their ice binding and ice nucleating properties. Amino acids, such as phenylalanine, leucine, isoleucine, and cysteine, were enriched in the SML [7,14]. Peptides containing some of these amino acids are known to assemble into hydrogels [159] that may make them good ice nuclei (IN) [160], and are thus important to be accounted for as controllers of high Arctic mixed phase clouds. In fact, many polar planktonic microbes, such as the diatom *Fragilariopsis cylindrus* [161] and the prokaryote *Colwellia*, are known to express antifreeze proteins [162]. Identifying the amino acid sequences that provide the hydrophobic and the hydrophilic surface-active properties of the proteins in the gel supramolecular assembly would thus elucidate a critical step that can alter cloud reflectivity. Furthermore, protein to carbohydrate ratio has been shown to be directly related to the relative hydrophobicity or "stickiness" of the EPS [163], thus protein enriched microgels at the air-water interface could provide a "glue" in the aerosols and thus, aid in the CCN formation.

Although airborne nanogels may have the chemical surfactant properties necessary to act as CCN, to behave as effective CCN they must first reach a critical size (≥50 nm) and meet physicochemical properties and energy constraints of the system [20,46]. Leck and Bigg [37,59] and Karl et al. [61] speculated that the primary marine gels would fragment [46] and disintegrate under some circumstances, generating progressively smaller particles, most likely due to UV radiation cleavage [62]. That this happened is suggested by the similarity in the shape of the size distributions of air and water [15] with microcolloidal size aggregates < 70 nm diameter. The measured modal diameters were 30 and 50 nm, respectively, shifting the airborne distribution to smaller sizes and being consistent with the hypothesis that fragmented gel aggregates may form almost all the aerosol particles between ca. 15 and 80 nm diameter [59]. Indeed, on average, during five weeks spent in the high Arctic pack ice region during 2001 [46,57,59], surface microlayer-derived particles represented more than half the collected airborne submicron particles and, on all days, dominated the aerosol population below 70 nm diameter. The fragmentation of marine gel particles is a process that may also be governed by repeated condensation and dissipation of fogs or clouds, following the strong indication of fog-related aerosol source mentioned in Heitzenberg et al. [164].

However, it is not yet clear whether the polymeric material reached smaller sizes due to cleavage caused by UV radiation, or due to reversible volume phase transition induced by poorly understood stimuli at this time, due to dispersion of the gels during their time travelling as aerosols, and/or perhaps by a combination of all three factors, as has been described for other marine gels [5,7,62]; thus, this remains an open research question. Furthermore, atmospheric gels could also provide sites for condensation of the biogenic gas DMS and/or its oxidation products, which can form an appreciable part of the total high Arctic aerosol [156]. In 2005, Leck and Bigg [15] detected marine organic—presumably microgel—material in half or more of their aerosol samples, which were also coated with sulfuric acid.

The co-occurrence of atmospheric organic material and biologically active marine waters has, since the mid-1990s, been confirmed for the high Arctic, but it has also now been documented for temperate waters [18,37,47,165], more recently reviewed by Quinn et al. [19]. In the emerging picture of the Arctic atmosphere, DMS concentrations will determine the mass of the particles by producing material for their growth. However, it is the number of airborne gels, or primary particles that will primarily influence the number of CCN and the resulting optical properties of the cloud droplets. Indeed, research during the past two decades—reviewed by Leck and Bigg [58] followed by Quinn and Bates [41]—does not corroborate the CLAW hypothesis for remote marine regions.

The transport of marine precursors to primary aerosols is usually a function of sea-spray, resulting from wind generated bubbles, or other bubble sources, bursting at the air–sea interface [35,42,130]. The presence of sea-ice in the central Arctic Ocean has long deterred and/or minimized the presence of sea spray. Indeed, Leck and Bigg [59] and Leck et al. [13,45] had hypothesized that the source of gels found in clouds was the open water between ice floes and that those particles would be transferred to the atmosphere by the bursting of air bubbles at the air–sea interface. However, it was only in 2011 that bubbles in the central Arctic Ocean were found at lower wind conditions (wind speeds $< 5 \text{ m s}^{-1}$) during clear sky days and ice melting to have similar size spectra to other oceans, although smaller sizes predominate [166], and provided a likely mechanism for getting SML-material airborne. The particles were mainly organic in nature, but their sea-salt component increased at high wind speeds $> 12 \text{ m s}^{-1}$ and dominated both mass and number. Interestingly, Nilsson et al. [167] showed that particle flux from the leads coincided with days of low winds, which also included many of the days with nucleation. It is now known that the diatom *M. arctica*'s "mucous matrix" traps oxygen bubbles produced during photosynthesis [168]. Thus, one can anticipate that these bubbles may burst, transferring biopolymeric material into the atmosphere. Since *M. arctica* is a ubiquitous diatom, forming 3–80% of the total phototrophic biomass in Arctic waters and being the most productive phytoplankton accounting for 90% of the primary production in the region [168], such oxygen bubble bursting from the mucous matrix of *M. arctica* may be an important, yet seldomly a quantified process [168]. However, the source(s) and abundance of these bubbles remain elusive to determine; during the most recent icebreaker expedition in summer of 2018 Arctic Ocean MOCCHA (Microbiology-Ocean-Cloud-Coupling in the High Arctic), the unexpected presence of small cap waves within the leads appear to dominate as a source of bubbles with respect to other possible sources (e.g., release from sea ice, photosynthesis, upward transport) [H. Czerki, pers. comm.].

8. Conclusions: Gels as a Source of CCN

The high Arctic studies of primary aerosol organic polymer gels derived from the SML between sea ice leads, performed over the last two decades, have been expanded to other marine areas [18,19,37,165,169]. These studies show strong similarities in the morphological and chemical characteristics of the particles in all regions [15,21,47]. Most of the studies have focused on size-segregated chemical characterization of marine aerosol particles [18,35,39,42,129,132,170,171] that, more recently, were reviewed by Quinn et al. [19,172]. They all have suggested marine DOC at the ocean surface to be the source for these aerosol particles. However, they have not examined nor quantified gel-specific, physical-chemical characteristics and their emergent properties (assembly, phase transition, and the effect of physico-chemical characteristics of the environment that determine the gel's pathway during assembly) of this sort of dissolved-to-particulate organic matter. Because this material has been confirmed to be DOC-based, these hydrated polymeric particles must be gels [5], and their emergent properties can provide insights into the processes controlling cloud formation, linking biology at the ocean surface with cloud properties. However, climate change and ocean acidification will increase temperature and lower pH that may, in turn, synergistically reduce the assembly of DOC polymers, as shown experimentally at temperatures applicable to tropical and subtropical areas [173], with significant effects on carbon cycling in the oceans, as well as affecting the production of aerosols and resultant cloud formation. We expect changes in the yield of the assembly in the Arctic as well, but this needs to be determined.

Gel-specific characteristics provide physico-chemical processes by which such particles can change their size spectra in a way that, when linked to aerosolization functions (i.e., sea spray, wind speed, sea surface temperature) (e.g., Salter et al. [174]) and marine DOC concentrations, would allow their parameterization into climate models.

Author Contributions: Conceptualization, M.V.O., D.A.H., P.A.M. and C.L.; writing original draft M.V.O.; Writing, review and editing, all authors. All authors have read and agreed to the published version of the manuscript.

Funding: This work was funded by NSF-OCE 1634009 to MVO, NSF-OCE 1436748 to DAH, NSF OPP-1724585 to PAM, and the Swedish Research Council (project no. 2016-03518) to CL.

Institutional Review Board Statement: Not applicable.

Informed Consent Statement: Not applicable.

Acknowledgments: We thank Pedro Verdugo (Dept. of Bioengineering, and Friday Harbor, Marine Laboratories, University of Washington) for his kind invitation to contribute to this special issue on marine gels.

Conflicts of Interest: The authors declare no conflict of interest. The funders had no role in the design of this study, or the writing of the manuscript or publishing this review.

References

1. Hansell, D.A.; Carlson, C.A.; Repeta, D.J.; Schlitzer, R. Dissolved organic matter in the ocean. A controversy stimulates new insights. *Oceanography* **2009**, *22*, 52–61. [CrossRef]
2. Siegenthaler, U.; Sarmiento, J.L. Atmospheric carbon dioxide and the ocean. *Nature* **1993**, *365*, 119–125. [CrossRef]
3. Hedges, J.I.; Keil, R.G. Sedimentary organic matter preservation: An assessment and speculative synthesis. *Mar. Chem.* **1995**, *49*, 81–115. [CrossRef]
4. Repeta, D.J. Chemical Characterization and Cycling of Dissolved Organic Matter. In *Biogeochemistry of Marine Dissolved Organic Matter*; Hansell, D.A., Carlson, C.A., Eds.; Elsevier: London, UK, 2015; p. 693.
5. Chin, W.-C.; Orellana, M.V.; Verdugo, P. Spontaneous assembly of marine dissolved organic matter into polymer gels. *Nature* **1998**, *391*, 568–572. [CrossRef]
6. Ding, Y.-X.; Chin, W.-C.; Rodriguez, A.; Hung, C.-C.; Santschi, H.P.; Verdugo, P. Amphiphilic exopolymers from Sagittula stellata induce DOM self-assembly and formation of marine microgels. *Mar. Chem.* **2008**, *112*, 11–19. [CrossRef]
7. Orellana, M.V.; Matrai, P.A.; Leck, C.; Rauschenberg, C.D.; Lee, A.M.; Coz, E. Marine microgels as a source of cloud condensation nuclei in the high Arctic. *Proc. Natl. Acad. Sci. USA* **2011**, *108*, 13612–13617. [CrossRef]
8. Radić, T.M.; Svetličić, V.; Žutić, V.; Boulgaropoulos, B. Seawater at the nanoscale: Marine gel imaged by atomic force microscopy. *J. Mol. Recognit.* **2011**, *24*, 397–405. [CrossRef]
9. Verdugo, P. Marine Microgels. *Ann. Rev. Mar. Sci.* **2012**, *4*, 375–400. [CrossRef]
10. Verdugo, P. Marine Biopolymer Dynamics, Gel Formation, and Carbon Cycling in the Ocean. *Gels* **2021**, *7*, 136. [CrossRef]
11. Orellana, M.V.; Leck, C. Marine Microgels. In *Biogeochemistry of Marine Dissolved Organic Matter*, 2nd ed.; Hansell, D.A., Carlson, C.A., Eds.; Academic Press: Boston, MA, USA, 2015; pp. 451–480.
12. Verdugo, P.; Santschi, P.H. Polymer dynamics of DOC networks and gel formation in seawater. *Deep. Sea Res. Part II Top. Stud. Oceanogr.* **2010**, *57*, 1486–1493. [CrossRef]
13. Bigg, E.K.; Leck, C. The composition of fragments of bubbles bursting at the ocean surface. *J. Geophys. Res.* **2008**, *113*, D11209. [CrossRef]
14. Matrai, P.A.; Tranvik, L.; Leck, C.; Knulst, J. Are high Arctic microlayers a potential source of aerosol organic precursors? *Mar. Chem.* **2008**, *108*, 109–122. [CrossRef]

15. Leck, C.; Bigg, E.K. Biogenic particles in the surface microlayer and overlaying atmosphere in the central Arctic Ocean during summer. *Tellus B Chem. Phys. Meteorol.* **2005**, *57*, 305–316. [CrossRef]
16. Ovadnevaite, J.; Ceburnis, D.; Martucci, G.; Bialek, J.; Monahan, C.; Rinaldi, M.; Facchini, M.C.; Berresheim, H.; Worsnop, D.R.; O'Dowd, C. Primary marine organic aerosol: A dichotomy of low hygroscopicity and high CCN activity. *Geophys. Res. Lett.* **2011**, *38*, L21806. [CrossRef]
17. Deng, C.; Brooks, S.D.; Vidaurre, G.; Thornton, D.C.O. Using Raman Microspectroscopy to Determine Chemical Composition and Mixing State of Airborne Marine Aerosols over the Pacific Ocean. *Aerosol Sci. Technol.* **2014**, *48*, 193–206. [CrossRef]
18. Russell, L.M.; Hawkins, L.N.; Frossard, A.A.; Quinn, P.K.; Bates, T.S. Carbohydrate-like composition of submicron atmospheric particles and their production from ocean bubble bursting. *Proc. Natl. Acad. Sci. USA* **2010**, *107*, 6652–6657. [CrossRef] [PubMed]
19. Quinn, P.K.; Collins, D.B.; Grassian, V.H.; Prather, K.A.; Bates, T.S. Chemistry and Related Properties of Freshly Emitted Sea Spray Aerosol. *Chem. Rev.* **2015**, *115*, 4383–4399. [CrossRef]
20. Leck, C.; Svensson, E. Importance of aerosol composition and mixing state for cloud droplet activation over the Arctic pack ice in summer. *Atmos. Chem. Phys.* **2015**, *15*, 2545–2568. [CrossRef]
21. Leck, C.; Gao, G.; Rad, M.F.; Nilsson, U. Size resolved airborne particulate polysaccharides in summer high. *Atmos. Chem. Phys. Discuss.* **2013**, *13*, 9801–9847.
22. O'Dowd, C.; Ceburnis, D.; Ovadnevaite, J.; Bialek, J.; Stengel, D.B.; Zacharias, M.; Nitschke, U.; Connan, S.; Rinaldi, M.; Fuzzi, S.; et al. Connecting marine productivity to sea-spray via nanoscale biological processes: Phytoplankton Dance or Death Disco? *Sci. Rep.* **2015**, *5*, 14883. [CrossRef]
23. Stocker, T.F.; Qin, D.; Plattner, G.-K.; Tignor, M.; Allen, S.K.; Boschung, J.; Nauels, A.; Xia, Y.; Bex, V.; Midgley, P.M. (Eds.) IPCC, 2013: Summary for Policymakers. In *Climate Change 2013: The Physical Science Basis. Contribution of Working Group I to the Fifth Assessment Report of the Intergovernmental Panel on Climate Change*; Intergovernmental Panel on Climate Change (IPCC): Cambridge, UK; New York, NY, USA, 2013; p. 30.
24. IPCC. *IPCC Presents Findings of the Special Report on Global Warming of 1.5 °C at Event to Discuss Viet Nam's Response to Climate Change*; Press Release: Geneva, Switzeland, 2018.
25. Twomey, S. Aerosols, clouds and radiation. *Atmos. Environ. Part A Gen. Top.* **1991**, *25*, 2435–2442. [CrossRef]
26. Mauritsen, T.; Sedlar, J.; Tjernström, M.; Leck, C.; Martin, M.; Shupe, M.; Sjogren, S.; Sierau, B.; Persson, P.O.G.; Brooks, I.M.; et al. An Arctic CCN-limited cloud-aerosol regime. *Atmos. Chem. Phys.* **2011**, *11*, 165–173. [CrossRef]
27. Twomey, S. The Influence of Pollution on the Shortwave Albedo of Clouds. *J. Atmos. Sci.* **1977**, *34*, 1149–1152. [CrossRef]
28. Bigg, E.K.; Leck, C. Cloud-active particles over the central Arctic Ocean. *J. Geophys. Res. Atmos.* **2001**, *106*, 32155–32166. [CrossRef]
29. Leck, C.; Persson, C. Seasonal and short-term variability in dimethyl sulfide, sulfur dioxide and biogenic sulfur and sea salt aerosol particles in the arctic marine boundary layer during summer and autumn. *Tellus Ser. B Chem. Phys. Meteorol.* **1996**, *48B*, 272–299. [CrossRef]
30. Chang, R.Y.W.; Leck, C.; Graus, M.; Müller, M.; Paatero, J.; Burkhart, J.F.; Stohl, A.; Orr, L.H.; Hayden, K.; Li, S.M.; et al. Aerosol composition and sources in the central Arctic Ocean during ASCOS. *Atmos. Chem. Phys.* **2011**, *11*, 10619–10636. [CrossRef]
31. Leck, C.; Tjernström, M.; Matrai, P.; Swietlicki, E.; Bigg, K. Can marine micro-organisms influence melting of the Arctic pack-ice? *EOS Trans. AGU* **2004**, *85*, 25–32. [CrossRef]
32. Duce, R.A.; Hoffman, E.J. Chemical fractionation at the air/sea interface. *Ann. Rev. Earth Planet. Sci.* **1976**, *4*, 187–228. [CrossRef]
33. Facchini, M.C.; Decesari, S.; Rinaldi, M.; Carbone, C.; Finessi, E.; Mircea, M.; Fuzzi, S.; Moretti, F.; Tagliavini, E.; Ceburnis, D.; et al. Important source of marine secondary organic aerosol from biogenic amines. *Environ. Sci. Technol.* **2008**, *42*, 9116–9121. [CrossRef] [PubMed]
34. Gaston, C.J.; Furutani, H.; Guazzotti, S.A.; Coffee, K.R.; Bates, T.S.; Quinn, P.K.; Aluwihare, L.I.; Mitchell, B.G.; Prather, K.A. Unique ocean-derived particles serve as a proxy for changes in ocean chemistry. *J. Geophys. Res. Atmos.* **2011**, *116*, D18310. [CrossRef]
35. Keene, W.C.; Maring, H.; Maben, J.R.; Kieber, D.J.; Pszenny, A.A.P.; Dahl, E.E.; Izaguirre, M.A.; Davis, A.J.; Long, M.S.; Zhou, X.; et al. Chemical and physical characteristics of nascent aerosols produced by bursting bubbles at a model air-sea interface. *J. Geophys. Res.* **2007**, *112*, D21202. [CrossRef]
36. Leck, C.; Bigg, E.K. Source and evolution of the marine aerosol—A new perspective. *Geophys. Res. Lett.* **2005**, *32*, L19803. [CrossRef]
37. Leck, C.; Bigg, E.K. New Particle Formation of Marine Biological Origin. *Aerosol Sci. Technol.* **2010**, *44*, 570–577. [CrossRef]
38. Middlebrook, A.M.; DMurphy, M.; Thomson, D.S. Observations of organic material in individual marine particles at Cape Grim during the First Aerosol Characterization Experiment (ACE 1). *J. Geophys. Res. Atmos.* **1998**, *103*, 16475–16483. [CrossRef]
39. O'Dowd, C.D.; Facchini, M.C.; Cavalli, F.; Ceburnis, D.; Mircea, M.; Decesari, S.; Fuzzi, S.; Yoon, Y.J.; Putaud, J.-P. Biogenically driven organic contribution to marine aerosol. *Nature* **2004**, *431*, 676–680. [CrossRef]
40. Yoon, Y.J.; Ceburnis, D.; Cavalli, F.; Jourdan, O.; Putaud, J.P.; Facchini, M.C.; Decesari, S.; Fuzzi, S.; Sellegri, K.; Jennings, S.G.; et al. Seasonal characteristics of the physicochemical properties of North Atlantic marine atmospheric aerosols. *J. Geophys. Res. Atmos.* **2007**, *112*, D04206. [CrossRef]
41. Quinn, P.K.; Bates, T.S. The case against climate regulation via oceanic phytoplankton sulphur emissions. *Nature* **2011**, *480*, 51–56. [CrossRef] [PubMed]

42. Quinn, P.K.; Bates, T.S.; Schulz, K.S.; Coffman, D.J.; Frossard, A.A.; Russell, L.M.; Keene, W.C.; Kieber, D.J. Contribution of sea surface carbon pool to organic matter enrichment in sea spray aerosol. *Nat. Geosci.* **2014**, *7*, 228–232. [CrossRef]
43. Dueker, M.E.; O'Mullan, G.D.; Weathers, K.C.; Juhl, A.R.; Uriarte, M. Coupling of fog and marine microbial content in the near-shore coastal environment. *Biogeosciences* **2012**, *9*, 803–813. [CrossRef]
44. Fröhlich-Nowoisky, J.; Kampf, C.J.; Weber, B.; Huffman, J.A.; Pöhlker, C.; Andreae, M.O.; Lang-Yona, N.; Burrows, S.M.; Gunthe, S.S.; Elbert, W.; et al. Bioaerosols in the Earth system: Climate, health, and ecosystem interactions. *Atmos. Res.* **2016**, *182*, 346–376. [CrossRef]
45. Leck, C.; Norman, M.; Bigg, E.K.; Hillamo, R. Chemical composition and sources of the high Arctic aerosol relevant for cloud formation. *J. Geophys. Res.* **2002**, *107*, 4135. [CrossRef]
46. Sun, J.; Ariya, P.A. Atmospheric organic and bio-aerosols as cloud condensation nuclei (CCN): A review. *Atmos. Environ.* **2006**, *40*, 795–820. [CrossRef]
47. Leck, C.; Bigg, E.K. Comparison of sources and nature of the tropical aerosol with the summer high Arctic aerosol. *Tellus B* **2008**, *60*, 118–126. [CrossRef]
48. Bigg, E.K.; Leck, C.; Tranvik, L. Particulates of the surface microlayer of open water in the central Arctic Ocean in summer. *Mar. Chem.* **2004**, *91*, 131–141. [CrossRef]
49. Alldredge, A.L.; Passow, U.; Logan, E.B. The abundance and significance of a class of large, transparent organic particles in the oceans. *Deep-Sea Res.* **1993**, *40*, 1131–1140. [CrossRef]
50. Passow, U.; Alldredge, A.L. A dye-binding assay for the spectrophotometric measurement of transparent exopolymer particles (TEP) in the ocean. *Limnol. Oceanogr.* **1995**, *40*, 1326–1335. [CrossRef]
51. Mopper, K.; Zhou, J.; Ramana, K.S.; Passow, U.; Dam, H.G.; Drapeau, D.T. The role of surface-active carbohydrates in the flocculation of a diatom bloom in a mesocosm. *Deep Sea Res. Part II Top. Stud. Oceanogr.* **1995**, *42*, 47–73. [CrossRef]
52. Hung, C.-C.; Tang, D.; Warnken, K.; Santschi, P.H. Distributions of carbohydrates, including uronic acids, in estuarine waters of Galveston Bay. *Mar. Chem.* **2001**, *73*, 305–318. [CrossRef]
53. Hung, C.-C.; Guo, L.; Schultz, G.E.; Pinckney, J.L.; Santschi, P.H. Production and flux of carbohydrate species in the Gulf of Mexico. *Glob. Biogeochem. Cycles* **2003**, *17*, 1055. [CrossRef]
54. Cunliffe, M.; Engel, A.; Frka, S.; Gašparović, B.; Guitart, C.; Murrell, J.C.; Salter, M.; Stolle, C.; Upstill-Goddard, R.; Wurl, O. Sea surface microlayers: A unified physicochemical and biological perspective of the air–ocean interface. *Prog. Oceanogr.* **2013**, *109*, 104–116. [CrossRef]
55. Aller, J.Y.; Radway, J.C.; Kilthau, W.P.; Bothe, D.W.; Wilson, T.W.; Vaillancourt, R.D.; Quinn, P.K.; Coffman, D.J.; Murray, B.J.; Knopf, D.A. Size-resolved characterization of the polysaccharidic and proteinaceous components of sea spray aerosol. *Atmos. Environ.* **2017**, *154*, 331–347. [CrossRef]
56. Charlson, R.J.; Lovelock, J.E.; Andreae, M.O.; Warren, S.G. Oceanic phytoplankton, atmospheric sulfur, cloud albedo and climate. *Nature* **1987**, *326*, 655–661. [CrossRef]
57. Leck, C.; Persson, C. The central Arctic as a source of dimethyl sulfide-Seasonal variability in relation to biological activity. *Tellus B* **1996**, *48*, 156–177. [CrossRef]
58. Leck, C.; Bigg, E.K. A modified aerosol–cloud–climate feedback hypothesis. *Environm. Chem.* **2007**, *4*, 400–403. [CrossRef]
59. Leck, C.; Bigg, E. Aerosol production over remote marine areas—A new route. *Geophys. Res. Lett.* **1999**, *23*, 23. [CrossRef]
60. Tjernstrom, M.; Leck, C.; Birch, C.E.; Brooks, B.J.; Brooks, I.M.; BÃ¤cklin, L.; Chang, R.Y.-W.; Granath, E.; Graus, M.; Hansel, A.; et al. The Arctic Summer Cloud-Ocean Study (ASCOS): Overview and experimental design. *Atmos. Chem. Phys. Discuss.* **2013**, *13*, 13541–13652.
61. Karl, M.; Leck, C.; Coz, E.; Heintzenberg, J. Marine nanogels as a source of atmospheric nanoparticles in the high Arctic. *Geophys. Res. Lett.* **2013**, *40*, 3738–3743. [CrossRef]
62. Orellana, M.V.; Verdugo, P. Ultraviolet radiation blocks the organic carbon exchange between the dissolved phase and the gel phase in the ocean. *Limnol. Oceanogr.* **2003**, *48*, 1618–1623. [CrossRef]
63. Orellana, M.V.; Matrai, P.A.; Janer, M.; Rauschenberg, C. DMSP storage in Phaeocystis secretory vesicles. *J. Phycol.* **2011**, *47*, 112–117. [CrossRef]
64. Martin, M.; Chang, R.Y.W.; Sierau, B.; Sjogren, S.; Swietlicki, E.; Abbatt, J.P.D.; Leck, C.; Lohmann, U. Cloud condensation nuclei closure study on summer arctic aerosol. *Atmos. Chem. Phys.* **2011**, *11*, 11335–11350. [CrossRef]
65. Prisle, N.L.; Raatikainen, T.; Laaksonen, A.; Bilde, M. Surfactants in cloud droplet activation: Mixed organic-inorganic particles. *Atmos. Chem. Phys.* **2010**, *10*, 5663–5683. [CrossRef]
66. Ovadnevaite, J.; Zuend, A.; Laaksonen, A.; Sanchez, K.J.; Roberts, G.; Ceburnis, D.; Decesari, S.; Rinaldi, M.; Hodas, N.; Facchini, M.C.; et al. Surface tension prevails over solute effect in organic-influenced cloud droplet activation. *Nature* **2017**, *546*, 637–641. [CrossRef] [PubMed]
67. Forestieri, S.D.; Staudt, S.M.; Kuborn, T.M.; Faber, K.; Ruehl, C.R.; Bertram, T.H.; Cappa, C.D. Establishing the impact of model surfactants on cloud condensation nuclei activity of sea spray aerosol mimics. *Atmos. Chem. Phys.* **2018**, *18*, 10985–11005. [CrossRef]
68. Sierau, B.; Chang, R.Y.W.; Leck, C.; Paatero, J.; Lohmann, U. Single-particle characterization of the high-Arctic summertime aerosol. *Atmos. Chem. Phys.* **2014**, *14*, 7409–7430. [CrossRef]
69. Hamacher-Barth, E.; Leck, C.; Jansson, K. Size-resolved morphological properties of the high Arctic summer aerosol during ASCOS-2008. *Atmos. Chem. Phys.* **2016**, *16*, 6577–6593. [CrossRef]

70. Baccarini, A.; Karlsson, L.; Dommen, J.; Duplessis, P.; Vüllers, J.; Brooks, I.M.; Saiz-Lopez, A.; Salter, M.; Tjernström, M.; Baltensperger, U.; et al. Frequent new particle formation over the high Arctic pack ice by enhanced iodine emissions. *Nat. Commun.* **2020**, *11*, 4924. [CrossRef] [PubMed]
71. Hill, V.L.; Manley, S.L. Release of reactive bromine and iodine from diatoms and its possible role in halogen transfer in polar and tropical oceans. *Limnol. Oceanogr.* **2009**, *54*, 812–822. [CrossRef]
72. Xu, C.; EMiller, J.; Zhang, S.; Li, H.-P.; Ho, Y.-F.; Schwehr, K.A.; Kaplan, D.I.; Otosaka, S.; Roberts, K.A.; Brinkmeyer, R.; et al. Sequestration and Remobilization of Radioiodine (129I) by Soil Organic Matter and Possible Consequences of the Remedial Action at Savannah River Site. *Environ. Sci. Technol.* **2011**, *45*, 9975–9983. [CrossRef]
73. Schwehr, K.A.; Santschi, P.H. Sensitive determination of iodine species, including organo-iodine, for freshwater and seawater samples using high performance liquid chromatography and spectrophotometric detection. *Anal. Chim. Acta* **2003**, *482*, 59–71. [CrossRef]
74. Nguyen, Q.T.; Kristensen, T.B.; Hansen, A.M.K.; Skov, H.; Bossi, R.; Massling, A.; Sørensen, L.L.; Bilde, M.; Glasius, M.; Nøjgaard, J.K. Characterization of humic-like substances in Arctic aerosols. *J. Geophys. Res. Atmos.* **2014**, *119*, 5011–5027. [CrossRef]
75. Bowman, J.S.; Deming, J.W. Elevated bacterial abundance and exopolymers in saline frost flowers and implications for atmospheric chemistry and micro-bial dispersal. *Geophys. Res. Lett.* **2010**, *37*, L13501. [CrossRef]
76. Aluwihare, L.I.; Repeta, D.J.; Pantoja, S.; Johnson, C.G. Two chemically distinct pools of organic nitrogen accumulate in the ocean. *Science* **2005**, *308*, 1007–1010. [CrossRef] [PubMed]
77. McCarthy, M.D.; Hedges, J.I.; Benner, R. Major bacterial contribution to marine dissolved organic nitrogen. *Science* **1998**, *281*, 231–234. [CrossRef] [PubMed]
78. Kaiser, K.; Benner, R. Major bacterial contribution to the ocean reservoir of detrital organic carbon and nitrogen. *Limnol. Oceanogr.* **2008**, *53*, 99–112. [CrossRef]
79. Hansell, D.A. Recalcitrant Dissolved Organic Carbon Fractions. In *Annual Review of Marine Science*; Carlson, C.A., Giovannoni, S.J., Eds.; Elsevier: Waltham, MA, USA, 2013; Volume 5, pp. 421–445.
80. Bauer, J.E.; Williams, P.M.; Druffel, E.R.M. C-14 Activity of dissolved organic carbon fractions in the North-central Pacific and Sargasso Sea. *Nature* **1992**, *357*, 667–670. [CrossRef]
81. Santschi, P.H.; Balnois, E.; Wilkinson, K.J.; Zhang, J.; Buffle, J.; Guo, L. Fibrillar polysaccharides in marine macromolecular organic matter as imaged by atomic force microscopy and transmission electron microscopy. *Limnol. Oceanogr.* **1998**, *43*, 896–908. [CrossRef]
82. Longnecker, K.; Kido Soule, M.C.; Kujawinski, E.B. Dissolved organic matter produced by Thalassiosira pseudonana. *Mar. Chem.* **2015**, *168*, 114–123. [CrossRef]
83. Carlson, C.A.; Hansell, D.A. DOM sources, sinks, reactivity and budgets. In *Biogeochemistry of Marine Dissolved Organic Matter*; Hansell, D.A., Carlson, C.A., Eds.; Elsevier: Waltham, MA, USA, 2015.
84. Decho, A.W. Microbial exopolymer secretions in ocean environments: Their role(s) in food webs and marine processes. *Oceanogr. Mar. Biol. Annu. Rev.* **1990**, *28*, 73–153.
85. Decho, A.W.; Gutierrez, T. Microbial Extracellular Polymeric Substances (EPSs) in Ocean Systems. *Front. Microbiol.* **2017**, *8*, 922. [CrossRef] [PubMed]
86. Proctor, L.M.; Fuhrman, J.A. Viral mortality of marine bacteria and cyanobacteria. *Nature* **1990**, *343*, 60–62. [CrossRef]
87. Suttle, C.A. Marine viruses—Major players in the global ecosystem. *Nat. Rev. Microbiol.* **2007**, *5*, 801–812. [CrossRef] [PubMed]
88. Berges, J.A.; Falkowski, P.G. Physiological stress and cell death in marine phytoplankton: Induction of proteases in response to nitrogen or light limitation. *Limnol. Oceangr.* **1998**, *43*, 129:135. [CrossRef]
89. Biddle, K. The Molecular Ecophysiology of Programmed Cell Death in Marine Phytoplankton. *Annu. Rev. Mar. Sci.* **2015**, *7*, 341–375. [CrossRef]
90. Orellana, M.V.; Pang, W.L.; Durand, P.M.; Whitehead, K.; Baliga, N.S. A Role for Programmed Cell Death in the Microbial Loop. *PLoS ONE* **2013**, *8*, e62595. [CrossRef] [PubMed]
91. Nagata, T.; Kirchman, D. Role of submicron particles and colloids in microbial food webs and biogeochemical cycles within marine environments. *Adv. Microb. Ecol.* **1997**, *15*, 81–103.
92. Strom, S.L. Microbial Ecology of Ocean Biogeochemistry: A Community Perspective. *Science* **2008**, *320*, 1043–1045. [CrossRef] [PubMed]
93. Strom, S.L.; Benner, R.; Ziegler, S.; Dagg, M.J. Planktonic grazers are a potentially important source of marine dissolved organic carbon. *Limnol. Oceangr.* **1997**, *42*, 1364–1374. [CrossRef]
94. Chin, W.-C.; Orellana, M.V.; Quesada, I.; Verdugo, P. Secretion in unicellular marine phytoplankton: Demonstration of regulated exocytosis in Phaeocystis globosa. *Plant Cell Physiol.* **2004**, *45*, 535–542. [CrossRef] [PubMed]
95. Aluwihare, L.I.; Repeta, D.J. A comparison of the chemical characteristics of oceanic DOM and extracellular DOM produced by marine algae. *Mar. Ecol. Prog. Ser.* **1999**, *186*, 105–117. [CrossRef]
96. Biddanda, B.; Benner, R. Carbon, nitrogen, and carbohydrate fluxes during the production of particulate and dissolved organic matter by marine phytoplankton. *Limnol. Oceanogr.* **1997**, *42*, 506–518. [CrossRef]
97. Aluwihare, L.I.; Repeta, D.J.; Chen, R.F. A major biopolymeric component to dissolved organic carbon in surface sea water. *Nature* **1997**, *387*, 166–169. [CrossRef]

98. Orellana, M.V.; Petersen, T.W.; Diercks, A.H.; Donohoe, S.; Verdugo, P.; van den Engh, G. Marine microgels: Optical and proteomic fingerprints. *Mar. Chem.* **2007**, *105*, 229–239. [CrossRef]
99. Popendorf, K.J.; Lomas, M.W.; van Mooy, B.A.S. Microbial sources of intact polar diacylglycerolipids in the western North Atlantic Ocean. *Org. Geochem.* **2011**, *42*, 803–811. [CrossRef]
100. Wakeham, S.G.; Pease, T.K.; Benner, R. Hydroxy fatty acids in marine dissolved organic matter as indicators of bacterial membrane material. *Org. Geochem.* **2003**, *34*, 857–868. [CrossRef]
101. Kujawinski, E.B.; Longnecker, K.; Blough, N.V.; Vecchio, R.D.; Finlay, L.; Kitner, J.B.; Giovannoni, S.J. Identification of possible source markers in marine dissolved organic matter using ultrahigh resolution mass spectrometry. *Geochim. Cosmochim. Acta* **2009**, *73*, 4384–4399. [CrossRef]
102. Kujawinski, E.B.; Giovannoni, S.; Longernecker, K.; MacDonald, J.; Kitner, J.B. The role of Sar 11 controling the molecular level composition of marine dissolved organic matter. In Proceedings of the 13th Internaitonal Symphosium on Microbial Ecology: Microbes—Stewards of Changing Planet 2010, Seattle, WA, USA, 22–27 August 2010.
103. Azam, F. The ecological role of water-column microbes in the sea. *Mar. Ecol. Prog. Ser.* **1983**, *10*, 257–263. [CrossRef]
104. Azam, F.; Malfatti, F. Microbial structuring of marine ecosystems. *Nat. Rev. Microbiol.* **2007**, *5*, 782–791. [CrossRef]
105. Amon, R.M.W.; Benner, R. Bacterial utilization of different size classes of dissolved organic matter. *Limnol. Oceanogr.* **1996**, *41*, 41–51. [CrossRef]
106. Tanoue, E.; Nishiyama, S.; Kamo, M.; Tsugita, A. Bacterial membranes: Possible source of dissolved protein in seawater. *Geochim. Cosmochim. Acta* **1995**, *59*, 2643–2648. [CrossRef]
107. Tanoue, E.; Ischii, M.; Midorikawa, T. Discrete dissolved and particulate proteins in oceanic waters. *Limnol. Oceanogr.* **1996**, *41*, 1334–1343. [CrossRef]
108. Smith, D.C.; Simon, M.; Alldredge, A.L.; Azam, F. Intense hydrolytic enzyme activity on marine aggregates and implications for rapid particle dissolution. *Nature* **1992**, *359*, 139–142. [CrossRef]
109. Jiao, N.; Herndl, G.J.; Hansell, D.A.; Benner, R.; Kattner, G.; Wilhelm, S.W.; Kirchman, D.L.; Weinbauer, M.G.; Luo, T.; Chen, F.; et al. Microbial production of recalcitrant dissolved organic matter: Long-term carbon storage in the global ocean. *Nat. Rev. Microbiol.* **2010**, *8*, 593–599. [CrossRef]
110. Nagata, T.; Tamburini, C.; Arístegui, J.; Baltar, F.; Bochdansky, A.B.; Fonda-Umani, S.; Fukuda, H.; Gogou, A.; Hansell, D.A.; Hansman, R.L.; et al. Emerging concepts on microbial processes in the bathypelagic ocean—Ecology, biogeochemistry, and genomics. *Deep Sea Res. Part II Top. Stud. Oceanogr.* **2010**, *57*, 1519–1536. [CrossRef]
111. Matrai, P.A.; Vernet, M.; Hood, R.; Jennings, A.; Brody, E.; Saemundsdottir, S. Light-dependence of carbon and sulfur production by polar clones of the genus Phaeocystis. *Mar. Biol.* **1995**, *124*, 1157–1167. [CrossRef]
112. Mock, T.; Otillar, R.P.; Strauss, J.; McMullan, M.; Paajanen, P.; Schmutz, J.; Salamov, A.; Sanges, R.; Toseland, A.; Ward, B.J.; et al. Evolutionary genomics of the cold-adapted diatom Fragilariopsis cylindrus. *Nature* **2017**, *541*, 536–540. [CrossRef]
113. Mock, T.; Samanta, M.P.; Iverson, V.; Berthiaume, C.; Robison, M.; Holtermann, K.; Durkin, C.; BonDurant, S.S.; Richmond, K.; Rodesch, M.; et al. Whole-genome expression profiling of the marine diatom Thalassiosira pseudonana identifies genes involved in silicon bioproceses. *Proc. Natl Acad. Sci. USA* **2008**, *105*, 1579–1584. [CrossRef] [PubMed]
114. Janech, M.G.; Krell, A.; Mock, T.; Kang, J.-S.; Raymond, J.A. ICE-BINDING PROTEINS FROM SEA ICE DIATOMS (BACILLARIOPHYCEAE)1. *J. Phycol.* **2006**, *42*, 410–416. [CrossRef]
115. Blanc, G.; Agarkova, I.; Grimwood, J.; Kuo, A.; Brueggeman, A.; Dunigan, D.D.; Gurnon, J.; Ladunga, I.; Lindquist, E.; Lucas, S.; et al. The genome of the polar eukaryotic microalga Coccomyxa subellipsoidea reveals traits of cold adaptation. *Genome Biol.* **2012**, *13*, R39. [CrossRef]
116. Deming, J.W.; Young, J.N. The Role of Exopolysaccharides in Microbial Adaptation to Cold Habitats. In *Psychrophiles: From Biodiversity to Biotechnology*, 2nd ed.; Margesin, R., Ed.; Springer: Berlin/Heidelberg, Germany, 2017.
117. Krembs, C.; Eicken, H.; Deming, J.W. Exopolymer alteration of physical properties of sea ice and implications for ice habitability and biogeochemistry in a warmer Arctic. *Proc. Natl. Acad. Sci. USA* **2011**, *108*, 3653–3658. [CrossRef] [PubMed]
118. Liss, P.; Duce, R. *The Sea Surface and Global Change*; Cambridge University Press: Cambridge, UK, 2005.
119. Gao, Q.; Leck, C.; Rauschenberg, C.; Matrai, P.A. On the chemical dynamics of extracellular polysaccharides in the high Arctic surface microlaye. *Ocean Sci.* **2012**, *8*, 401–418. [CrossRef]
120. Wingender, J.; Neu, T.; Flemming, H.C. (Eds.) *Microbial Extracellular Polymeric Substances: Characterization, Structure and Function*; Springer Science and Business Media: Heidelberg, Germany, 1999.
121. Aller, J.Y.; Kuznetsova, M.R.; Jahns, C.J.; Kemp, P.F. The sea surface microlayer as a source of viral and bacterial enrichment in marine aerosols. *J. Aerosol Sci.* **2005**, *36*, 801–812. [CrossRef]
122. Blanchard, D.C. Bubble scavenging and the water to air transfer of organic material in the sea. *Adv. Chem. Ser.* **1976**, *145*, 360–387.
123. Blanchard, D.C.; Syzdek, L.D. Film drop production as a function of bubble size. *J. Geophys. Res.* **1988**, *93*, 3649–3654. [CrossRef]
124. Matrai, P.A.; Vernet, M.; Wassmann, P. Relating temporal and spatial patterns of DMSP in the Barents Sea to phytoplankton biomass and productivity. *J. Mar. Syst.* **2007**, *67*, 87–101. [CrossRef]
125. Ovadnevaite, J.; Ceburnis, D.; Leinert, S.; Dall'Osto, M.; Canagaratna, M.; O'Doherty, S.; Berresheim, H.; O'Dowd, C. Submicron NE Atlantic marine aerosol chemical composition and abundance: Seasonal trends and air mass categorization. *J. Geophys. Res. Atmos.* **2014**, *119*, 11850–11863. [CrossRef]

126. Vardi, A.; Haramaty, L.; van Mooy, B.A.S.; Fredricks, H.F.; Kimmance, S.A.; Larsen, A.; Bidle, K.D. Host–virus dynamics and subcellular controls of cell fate in a natural coccolithophore population. *Proc. Natl. Acad. Sci. USA* **2012**, *109*, 19327–19332. [CrossRef] [PubMed]
127. Rosenwasser, S.; Sheyn, U.; Frada, M.J.; Pilzer, D.; Rotkopf, R.; Vardi, A. Unmasking cellular response of a bloom-forming alga to viral infection by resolving expression profiles at a single-cell level. *PLoS Pathog.* **2019**, *15*, e1007708. [CrossRef]
128. Frossard, A.A.; Russell, L.M.; Massoli, P.; Bates, T.S.; Quinn, P.K. Side-by-Side Comparison of Four Techniques Explains the Apparent Differences in the Organic Composition of Generated and Ambient Marine Aerosol Particles. *Aerosol Sci. Technol.* **2014**, *48*, v–x. [CrossRef]
129. Frossard, A.A.; Russell, L.M.; Burrows, S.M.; Elliott, S.M.; Bates, T.S.; Quinn, P.K. Sources and composition of submicron organic mass in marine aerosol particles. *J. Geophys. Res. Atmos.* **2014**, *119*, 12977–13003. [CrossRef]
130. de Leeuw, G.; Andreas, E.L.; Anguelova, M.D.; Fairall, C.W.; Lewis, E.R.; O'Dowd, C.; Schulz, M.; Schwartz, S.E. Production flux of sea spray aerosol. *Rev. Geophys.* **2011**, *49*, RG2001. [CrossRef]
131. Long, M.S.; Keene, W.C.; Kieber, D.J.; Frossard, A.A.; Russell, L.M.; Maben, J.R.; Kinsey, J.D.; Quinn, P.K.; Bates, T.S. Light-enhanced primary marine aerosol production from biologically productive seawater. *Geophys. Res. Lett.* **2014**, *41*, 2661–2670. [CrossRef]
132. Cavalli, F. Advances in identification of organic matter in marine aerosol. *J. Geophys. Res.* **2004**, *109*. [CrossRef]
133. Beaupre, S.R.; Kieber, D.J.; Keene, W.C.; Long, M.S.; Maben, J.R.; Lu, X.; Zhu, Y.; Frossard, A.A.; Kinsey, J.D.; Bisgrove, J. Oceanic efflux of ancient marine dissolved organic carbon in primary marine aerosol. *Sci. Adv.* **2019**, *5*, eaax6535. [CrossRef] [PubMed]
134. Lawler, M.J.; Lewis, S.L.; Russell, L.M.; Quinn, P.K.; Bates, T.S.; Coffman, D.J.; Upchurch, L.M.; Saltzman, E.S. North Atlantic marine organic aerosol characterized by novel offline thermal desorption mass spectrometry: Polysaccharides, recalcitrant material, and secondary organics. *Atmos. Chem. Phys.* **2020**, *20*, 16007–16022. [CrossRef]
135. de Gennes, P.G.; Leger, L. Dynamics of entangled polymer chains. *Annu. Rev. Phys. Chem.* **1982**, *33*, 49–61. [CrossRef]
136. Edwards, S.F. The theory of macromolecular networks. *Biorheology* **1986**, *23*, 589–603. [CrossRef] [PubMed]
137. Li, X.; Leck, C.; Sun, L.; Hede, T.; Tu, Y.; Ågren, H. Cross-Linked Polysaccharide Assemblies in Marine Gels: An Atomistic Simulation. *J. Phys. Chem. Lett.* **2013**, *4*, 2637–2642. [CrossRef]
138. Frederick, J.E.; Snell, H.E.; Haywood, E.K. Solar ultraviolet radiation at the earth's surface. *Photochem. Photobiol.* **1989**, *50*, 443–450. [CrossRef]
139. Orellana, M.V.; Vetter, Y.A.; Verdugo, P. The assembly of DOM polymers into POM microgels enhances their suceptibility to bacterial degradation. In Proceedings of the Aquatic Sciences Meeting, San Antonio, TX, USA, 24–28 January 2000.
140. Sun, L.; Xu, C.; Lin, P.; Quigg, A.; Chin, W.-C.; Santschi, P.H. Photo-oxidation of proteins facilitates the preservation of high molecular weight dissolved organic nitrogen in the ocean. *Mar. Chem.* **2021**, *229*, 103907. [CrossRef]
141. Wesslén, C.; Tjernström, M.; Bromwich, D.H.; de Boer, G.; Ekman, A.M.L.; Bai, L.S.; Wang, S.H. The Arctic summer atmosphere: An evaluation of reanalyses using ASCOS data. *Atmos. Chem. Phys.* **2014**, *14*, 2605–2624. [CrossRef]
142. Kutschan, B.; Thoms, S.; Bayer-Giraldi, M. Thermal hysteresis of antifreeze proteins considering Fragilariopsis cylindrus. *Algol. Stud.* **2016**, *151–152*, 69–86. [CrossRef]
143. Tanaka, T.; Fillmore, D.; Sun, S.-T.; Nishio, I.; Swislow, G.; Shah, A. Phase transitions in ionic gels. *Phys. Rev. Lett.* **1980**, *45*, 1636–1639. [CrossRef]
144. Nishibori, N.; Matuyama, Y.; Uchida, T.; Moriyama, T.; Ogita, Y.; Oda, M.; Hirota, H. Spatial and temporal variations in free polyamine distributions in Uranouchi Inlet, Japan. *Mar. Chem.* **2003**, *82*, 307–314. [CrossRef]
145. Okajima, M.K.; Nguyen, Q.T.l.; Tateyama, S.; Masuyama, H.; Tanaka, T.; Mitsumata, T.; Kaneko, T. Photoshrinkage in Polysaccharide Gels with Trivalent Metal Ions. *Biomacromolecules* **2012**, *13*, 4158–4163. [CrossRef] [PubMed]
146. Rosen, S.L. *Fundamental Principles of Polymeric Materials*, 2nd ed.; Wiley and Sons, Inc.: New York, NY, USA, 1993; p. 420.
147. Wu, T.; Li, H. Liquid–solid phase transition of physical hydrogels subject to an externally applied electro-chemo-mechanical coupled field with mobile ionic species. *Phys. Chem. Chem. Phys.* **2017**, *19*, 21012–21023. [CrossRef]
148. Tanaka, T. *Phase Transitions of Gels, in Polyelectrolyte Gels*; American Chemical Society: Washington, DC, USA, 1992; pp. 1–21.
149. Dušek, K.; Patterson, D. Transition in swollen polymer networks induced by intramolecular condensation. *J. Polym. Sci. Part A-2 Polym. Phys.* **1998**, *6*, 1209–1216. [CrossRef]
150. Orellana, M.V.; Hansell, D. Ribulose-1,5-bisphosphate carboxylase/oxygenase (RuBisCO): A long-lived protein in the deep ocean. *Limnol. Oceanogr.* **2012**, *57*, 826–834. [CrossRef]
151. Kadko, D.; Galfond, B.; Landing, W.M.; Shelley, R.U. Determining the pathways, fate, and flux of atmospherically derived trace elements in the Arctic ocean/ice system. *Mar. Chem.* **2016**, *182*, 38–50. [CrossRef]
152. Paatero, J.; Vaattovaara, P.; Vestenius, M.; Meinander, O.; Makkonen, U.; Kivi, R.; Hyvärinen, A.; Asmi, E.; Tjernström, M.; Leck, C. Finnish contribution to the Arctic Summer Cloud Ocean Study (ASCOS) expedition, Arctic Ocean 2008. *Geophysica* **2009**, *45*, 119–146.
153. Maenhaut, W.; Ducastel, G.; Leck, C.C.; Nilsson, E.D.; Heintzenberg, J. Multi-elemental Composition and Sources of the High Arctic Atmospheric Aerosol during Summer and Autumn. *Tellus Ser. B—Chem. Phys. Meteorol.* **1996**, *48*, 300–321. [CrossRef]
154. Verdugo, P. Dynamics of marine biopolymer networks. *Polym. Bull.* **2007**, *58*, 139–143. [CrossRef]
155. Maitra, U.; Mukhpadhyay, S.; Sarkar, A.; Rao, P.; Indi, S.S. Hydrophobic pockets in a nonpolymeric aqueous gel: Observation of such a gelation process by color change. *Angew. Chem. Int. Ed.* **2001**, *40*, 2281–2283. [CrossRef]

156. Heintzenberg, J.; Leck, C.; Tunved, P. Potential source regions and processes of aerosol in the summer Arctic. *Atmos. Chem. Phys.* **2015**, *15*, 6487–6502. [CrossRef]
157. Lohmann, U.; Leck, C. Importance of submicron surface active organic aerosols for pristine Arctic clouds. *Tellus B* **2005**, *57*, 261–268. [CrossRef]
158. Bulatovic, I.; Igel, A.L.; Leck, C.; Heintzenberg, J.; Riipinen, I.; Ekman, A.M.L. The importance of Aitken mode aerosol particles for cloud sustenance in the summertime high Arctic—A simulation study supported by observational data. *Atmos. Chem. Phys.* **2021**, *21*, 3871–3897. [CrossRef]
159. Saiani, A.; Mohammed, A.; Frielinghaus, H.; Collins, R.; Hodson, N.; Kielty, C.M.; Sherratt, M.J.; Miller, A.F. Self-assembly and gelation properties of α-helix versus β-sheet forming peptides. *Soft Matter* **2009**, *5*, 193–202. [CrossRef]
160. Lorv, J.S.H.; Rose, D.R.; Glick, B.R. Bacterial Ice Crystal Controlling Proteins. *Scientifica* **2014**, *2014*, 976895. [CrossRef]
161. Bayer-Giraldi, M.; Weikusat, I.; Besir, H.; Dieckmann, G. Characterization of an antifreeze protein from the polar diatom Fragilariopsis cylindrus and its relevance in sea ice. *Cryobiology* **2011**, *63*, 210–219. [CrossRef] [PubMed]
162. Raymond, J.A.; Fritsen, C.; Shen, K. An ice-binding protein from an Antarctic sea ice bacterium. *FEMS Microbiol. Ecol.* **2007**, *61*, 214–221. [CrossRef] [PubMed]
163. Chen, C.-S.; Shiu, R.-F.; Hsieh, Y.-Y.; Xu, C.; Vazquez, C.I.; Cui, Y.; Hsu, I.C.; Quigg, A.; Santschi, P.H.; Chin, W.-C. Stickiness of extracellular polymeric substances on different surfaces via magnetic tweezers. *Sci. Total. Environ.* **2021**, *757*, 143766. [CrossRef]
164. Heintzenberg, J.; CLeck, C.; Birmili, W.; Wehner, B.; Tjernström, M.; Wiedensohler, A. Aerosol number–size distributions during clear and fog periods in the summer high Arctic: 1991, 1996 and 2001. *Tellus B Chem. Phys. Meteorol.* **2006**, *58*, 41–50. [CrossRef]
165. Facchini, M.C.; Rinaldi, M.; Decesari, S.; Carbone, C.; Finessi, E.; Mircea, M.; Fuzzi, S.; Ceburnis, D.; Flanagan, R.; Nilsson, E.D.; et al. Primary submicron marine aerosol dominated by insoluble organic colloids and aggregates. *Geophys. Res. Lett.* **2008**, *35*, L17814. [CrossRef]
166. Norris, S.J.; Brooks, I.M.; Leeuw, G.d.; Sirevaag, A.; Leck, C.; Brooks, B.J.; Birch, C.E.; Tjernstrom, M. Measurements of bubble size spectra within leads in the Arctic summer pack ice. *Ocean Sci. Discuss.* **2011**, *7*, 1739–1765. [CrossRef]
167. Nilsson, E.D.; Rannik, Ü.; Swietlicki, E.; Leck, C.; Aalto, P.P.; Zhou, J.; Norman, M. Turbulent aerosol fluxes over the Arctic Ocean: 2. Wind-driven sources from the sea. *J. Geophys. Res. Atmos.* **2001**, *106*, 32139–32154. [CrossRef]
168. Fernández-Méndez, M.; Wenzhöfer, F.; Peeken, I.; Sørensen, H.L.; Glud, R.N.; Boetius, A. Composition, Buoyancy Regulation and Fate of Ice Algal Aggregates in the Central Arctic Ocean. *PLoS ONE* **2014**, *9*, e107452. [CrossRef] [PubMed]
169. Fuentes, E.; Coe, H.; Green, D.; de Leeuw, G.; McFiggans, G. On the impacts of phytoplankton-derived organic matter on the properties of the primary marine aerosol—Part 1: Source fluxes. *Atmos. Chem. Phys.* **2010**, *10*, 9295–9317. [CrossRef]
170. Rinaldi, M.; Decesari, S.; Finessi, E.; Giulianelli, L.; Carbone, C.; Fuzzi, S.; O'Dowd, C.D.; Ceburnis, D.; Facchini, M.C. Primary and Secondary Organic Marine Aerosol and Oceanic Biological Activity: Recent Results and New Perspectives for Future Studies. *Adv. Meteorol.* **2010**, *2010*, 310682. [CrossRef]
171. Sciare, J.; Favez, O.; Sarda-Esteve, R.; Oikonomou, K.; Kazan, V. Long-term observations of carbonaceous aerosols in the Austral Ocean atmosphere: Evidence of a biogenic marine organic source. *J. Geophys. Res.* **2009**, *114*, D15302. [CrossRef]
172. Gantt, B.; Meskhidze, N. The physical and chemical characteristics of marine primary organic aerosol: A review. *Atmos. Chem. Phys.* **2013**, *13*, 3979–3996. [CrossRef]
173. Chen, C.S.; Anaya, J.M.; Chen, E.Y.; Farr, E.; Chin, W.C. Ocean warming-acidification synergism undermines dissolved organic matter assembly. *PLoS ONE* **2015**, *10*, e0118300. [CrossRef] [PubMed]
174. Salter, M.E.; Zieger, P.; Navarro, J.C.A.; Grythe, H.; Kirkevåg, A.; Rosati, B.; Riipinen, I.; Nilsson, E.D. An empirically derived inorganic sea spray source function incorporating sea surface temperature. *Atmos. Chem. Phys.* **2015**, *15*, 11047–11066. [CrossRef]

MDPI
St. Alban-Anlage 66
4052 Basel
Switzerland
Tel. +41 61 683 77 34
Fax +41 61 302 89 18
www.mdpi.com

Gels Editorial Office
E-mail: gels@mdpi.com
www.mdpi.com/journal/gels

www.ingramcontent.com/pod-product-compliance
Lightning Source LLC
LaVergne TN
LVHW070557100526
838202LV00012B/495
* 9 7 8 3 0 3 6 5 3 4 5 1 0 *